CAMBRIDGE MONOGRAPHS ON PHYSICS

GENERAL EDITORS

A. HERZENBERG, PH.D.
Reader in Theoretical Physics in the University of Manchester

J. M. ZIMAN, D.PHIL.
Professor of Theoretical Physics in the University of Bristol

SINGLE CRYSTAL DIFFRACTOMETRY

SINGLE CRYSTAL DIFFRACTOMETRY

BY

U . W . ARNDT

Medical Research Council
Laboratory of Molecular Biology, Cambridge

AND

B . T . M . WILLIS

Atomic Energy Research Establishment, Harwell

CAMBRIDGE
AT THE UNIVERSITY PRESS
1966

CAMBRIDGE UNIVERSITY PRESS
Cambridge, New York, Melbourne, Madrid, Cape Town, Singapore, São Paulo, Delhi

Cambridge University Press
The Edinburgh Building, Cambridge CB2 8RU, UK

Published in the United States of America by Cambridge University Press, New York

www.cambridge.org
Information on this title: www.cambridge.org/9780521112291

First published 1966
This digitally printed version 2009

A catalogue record for this publication is available from the British Library

Library of Congress Catalogue Card Number: 66–13637

ISBN 978-0-521-04060-0 hardback
ISBN 978-0-521-11229-1 paperback

To Valerie and Nan

CONTENTS

CHAPTER 4

Detectors

CHAPTER 5

Electronic Circuits

CHAPTER 6

The Production of the Primary Beam (X-rays)

CHAPTER 7

The Production of the Primary Beam (Neutrons)

CHAPTER 8

The Background

CHAPTER 12

Computer Programs and On-line Control

PREFACE

The determination of a crystal structure normally proceeds in three distinct stages. The first is the measurement of the intensities of the Bragg reflexions and the calculation from them of amplitudes, reduced to a common scale and corrected for various geometrical and physical factors. These amplitudes are known as 'observed structure amplitudes' or 'observed structure factors'. The second stage is the solution of the phase problem: the phases of the reflexions cannot be measured directly, and yet they must be derived in some way before the structure can be solved by Fourier methods. Because of uncertainties in the amplitudes and phases, this first structure is only approximately correct. The third stage in the structure determination consists of refining the approximate atomic positions so as to obtain the best possible agreement between the observed structure factors and the 'calculated structure factors', that is, those calculated from the approximate atomic positions of the successive stages of refinement.

This book describes counter methods of obtaining the set of observed structure factors of a single crystal, and so is concerned with the first stage only. The rapidly developing interest in automatic methods of collecting structure-factor data, and in measuring intensities to a high level of accuracy, have stimulated the development of counter methods. Photographic methods are still widely used in X-ray crystallography, but in neutron diffraction the single crystal diffractometer has remained the basic instrument for measuring neutron structure amplitudes since systematic studies began in the early 1950's. We shall cover, therefore, the fields of both X-ray and neutron diffractometry: there are remarkably few points where the two fields diverge and we discuss only one subject—the production of the primary beam—in separate chapters for X-rays and for neutrons. Comparisons between X-ray and neutron diffractometry are both interesting and illuminating, and many examples of such comparisons are given in the text.

We assume that the reader has a general acquaintance with the properties of the radiations with which we are concerned: these properties are described adequately in *X-rays in Theory and Experiment* by A. H. Compton and S. K. Allison, and in *Pile Neutron Research* by D. J. Hughes or *Neutron Diffraction* by G. E. Bacon. A familiarity with M. J. Buerger's *X-ray Crystallography*, a book which charted our own first steps in reciprocal space, and with F. C. Phillips's *An Introduction to Crystallography* would be very helpful.

Advances in single crystal diffractometry have been very rapid during the past few years, and some subjects which we discuss in this monograph are certain of further development. We anticipate, for instance, a better understanding of the various systematic errors occurring in the measurement of structure-factor data: the recent publication of the proceedings of a symposium of the American Crystallographic Association on 'Accuracy in X-ray Intensity Measurements' [Transactions of American Crystallographic Association, volume 1, 1965] testifies to the lively interest in this subject. We do not expect radically different geometrical or mechanical designs in diffractometers of the near future, but the on-line control of these instruments will become more common as cheaper small computers become available. In X-ray crystallography, diffractometers will have to compete in cost and simplicity of operation with computer-linked microdensitometers, and so there may be renewed interest in photographic methods. The eventual development of co-ordinate detectors, in both X-ray and neutron crystallography, could lead to a much more efficient data-collection procedure than any at present in use. For many years to come, however, it seems reasonably certain that automatic X-ray and neutron diffractometers will become standard instruments in an increasing number of laboratories.

We acknowledge the assistance generously given by many colleagues. Professor J. M. Ziman has made many editorial suggestions for improving the text. Individual chapters have been read by Drs D. M. Blow, E. Sandor and H. C. Watson, by Professor S. Ramaseshan, and by Mr J. F. Mallet, and many of their comments have been incorporated. We are especially indebted to Professor R. A. Young of Georgia State University, who has given us access to his extensive and unpublished writings on single crystal diffractometry. We are grateful to Valerie Arndt who prepared the index. We wish to thank the authors, publishers, and learned societies who have allowed us to reproduce published diagrams and photographs; the sources of these are given in the text. Finally, we should like to express our thanks to the publishers for their co-operation and patience.

U. W. A.
B. T. M. W.

April 1966

CHAPTER I

INTRODUCTION

Until recently, nearly all crystal-structure determinations with X-rays were carried out using photographic methods. Although the majority of crystal structures are still solved using photographic data, the past few years have witnessed profound changes in experimental techniques with the introduction of single crystal diffracton ters and with the application of automation procedures to their control. These changes, together with advances in crystallographic computing techniques, portend a sharp increase in both the quantity and quality of future crystallographic work.

In this chapter we shall sketch first the development of photographic and counter methods of measuring the set of X-ray intensities diffracted by a single crystal. This is followed by a survey of single crystal techniques in neutron diffraction and by a comparison of these techniques with X-ray methods. The chapter finishes with a discussion of automatic diffractometers and of the accuracy, speed and cost of counter methods. Our main aim in this chapter is to introduce briefly many points which are explained more fully later in the book.

1.1. X-ray techniques for measuring Bragg reflexions

The first diffraction pattern from a crystal, copper sulphate, was recorded on a photographic plate (Friedrich, Knipping & von Laue, 1912). Shortly afterwards, the ionization spectrometer (Bragg & Bragg, 1913) was developed and used both for the measurement of the wavelengths of X-ray spectra and for the determination of crystal structures. In order to examine the reflected X-ray beam with the spectrometer the crystal was suitably oriented and the intensity measured with an ionization chamber; however, because of the difficulty of measuring the very small ionization currents from an ionization chamber, the Bragg spectrometer was not really suited to the examination of the structures of small crystals or the collection of a large amount of intensity data, and during the 1920's the instrument was gradually superseded in structural work by the X-ray camera. The general acceptance of photographic methods was promoted especially by the application of the theory of the reciprocal lattice and the Ewald sphere of reflexion (Ewald, 1921) to the interpretation of single crystal rotation photographs (Bernal, 1926), and by the invention of the moving-film camera (Weissenberg, 1924); the Weissenberg camera permitted the measurement of the X-ray reflexions in a systematic way, covering one layer of reciprocal space at a time.

The principal advantage of photographic film over a counter detector is the possibility of recording a large number of reflexions on the same film; the principal drawback is the difficulty of relating the blackness of the film to the intensity of the diffracted beam producing it. For a long time the measurement of X-ray intensities was carried out by comparing visually the blackening of spots on the photographic film with a series of reference spots of graded intensity. The introduction of the precession camera (Buerger, 1942, 1944, 1964), giving an undistorted representation of the reciprocal lattice, encouraged the wider use of microdensitometry. Even now, however, there are very few microdensitometers in use which are capable of measuring automatically the integrated intensity of every spot on a precession photograph, let alone the intensities of spots on a Weissenberg photograph which lie along lines of non-uniform curvature.

Since 1945 the Weissenberg moving-film and the Buerger precession cameras have been the standard instruments for structural work, but during the same period interest has revived in counter methods. The Geiger counter and, later, the proportional and scintillation counters have been developed as reliable detectors, which supersede the ionization chamber used in the Bragg ionization spectrometer. These counters are quantum detectors, capable of counting individual X-ray quanta and of giving a more direct and accurate estimate of the diffracted intensity than photographic film. Extensive use has been made of these detectors for the direct recording of X-ray powder diffraction diagrams. Lonsdale (1948) and Cochran (1950) were amongst the first to exploit quantum counter methods in the examination of single crystals, and many later workers have acquired single crystal diffractometers to supplement or supplant the collection of diffraction data by X-ray cameras. In recent years there has been intense activity, especially in Europe and the U.S.A., devoted to developing automatic single crystal diffractometers, and a number of different kinds of these instruments is now available commercially.

We note here that, with the widespread use of quantum counters in diffraction work, the word 'diffractometer' was adopted in 1952 by the Apparatus Commission of the International Union of Crystallography to describe instruments for measuring diffracted X-rays (neutrons, electrons) by means of counter detectors. The word 'spectrometer' is reserved for instruments whose principal function is the investigation of X-ray wavelength (or neutron energy) spectra. Thus instruments used for X-ray fluorescence analysis, or the triple-axis instrument (Brockhouse, 1961) used to measure the energy transfer of neutrons scattered inelastically by crystals, are spectrometers. The Bragg ionization instrument would now be termed a diffractometer if used for structural work and a spectrometer for examining X-ray wavelength spectra.

1.2. Neutron techniques for measuring Bragg reflexions: comparison with X-ray techniques

Neutrons do not have a *direct* effect on photographic emulsions, and the diffractometer (using a proportional BF_3 counter as detector) has remained the basic instrument for measuring neutron structure factors. The first manual diffractometer designed for neutron work was built at the Argonne National Laboratory in the U.S.A. in 1945 (Zinn, 1947). During the following period of approximately twelve years nuclear research reactors were available for diffraction studies at only a few major centres: Oak Ridge (Wollan & Shull, 1948), Brookhaven (Corliss, Hastings & Brockman, 1953), Chalk River (Hurst, Pressessky & Tunnicliffe, 1950) and Harwell (Bacon, Smith & Whitehead, 1950). Many more research reactors have been constructed since the mid-1950's, and neutron diffractometers are now in operation at numerous nuclear research centres throughout the world.

In this book we are concerned mainly with those neutron techniques which have a counterpart in X-ray diffraction. For this reason, inelastic scattering methods, using either the triple-axis spectrometer or time-of-flight techniques, are largely ignored.

In the Appendix at the end of the book we list a number of differences between neutron and X-ray diffractometry: here we shall mention only two of the more obvious differences. The first arises from the nature of the source of the primary radiation. In X-ray crystallography, characteristic radiation of the target element is employed: this is in the form of sharp emission lines, rising

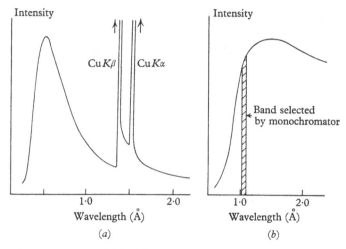

Fig. 1. Intensity curves: (a) for X-rays from a copper target which gives intense lines of characteristic K radiation; (b) for the slow neutron beam emerging from a reactor (after Bacon, 1962).

steeply from a background of 'white' or continuous radiation (Fig. 1a). The presence of white radiation can cause errors in estimating the contribution of the characteristic radiation to the Bragg reflexion. The white radiation can be removed with a single crystal monochromator, but this may lead to further difficulties, and frequently balanced filters or simple beta filters are used instead of crystal monochromators. In neutron diffraction, on the other hand, apart from a radically new technique which uses a pulsed neutron source and time-of-flight wavelength analysis (see p. 217), there is no real alternative to using a monochromator. The slow neutron spectrum from a nuclear reactor consists of a broad distribution of wavelengths: there is no sharply defined wavelength (Fig. 1*b*), and radiation from a narrow band of wavelengths is selected with a monochromator.

The second difference concerns the scale of the apparatus. With the highest flux reactors available at present the neutron flux at the sample, measured in neutrons/cm²/s, is always very much less than the flux, in X-ray quanta/cm²/s, from an X-ray tube. This flux difference requires the employment of larger samples in neutron diffraction. Moreover, materials for shielding against fast neutrons must be placed around the monochromator assembly and around the detector; fast neutron shielding is necessarily bulky and so neutron diffractometer assemblies tend to be larger than corresponding X-ray equipment.

1.3. Automatic diffractometers

There has always been a strong incentive to making neutron diffractometers as fully automatic as possible. The neutron source is much more expensive than an X-ray generator, and expenditure on a neutron diffractometer allowing utilization of the neutron source for 24 h/day is justified more readily on economic grounds. The development of automatic X-ray diffractometers has benefited much from the early application of automatic methods in the neutron field.

The setting of the crystal and detector shafts of a diffractometer, and the measurement of each Bragg reflexion, involve repetitive procedures which are extremely tedious to carry out by hand. Fig. 2 illustrates a simplified flow of operations in a system designed to undertake these procedures automatically. The input information consists of instructions for setting the crystal and the detector for each Bragg reflexion, and of instructions for measuring the intensity of the reflexion once the shafts are correctly set. Normally all reflexions are measured sequentially. The output data, consisting of the number of X-ray quanta or slow neutrons which are recorded in scanning across the reflexions, are presented in a form (for instance, punched paper tape) suitable for direct processing by a computer. In the 'off-line' system this processing leads directly to a set of structure amplitudes. In the 'on-line' method the computer is linked both to the diffractometer output and to its controls: by suitable programming, it is possible to scrutinize the experimental data as they are recorded and, if a measurement fails to satisfy certain programmed criteria, remedial action can be taken, consisting of automatic adjustment of the measuring

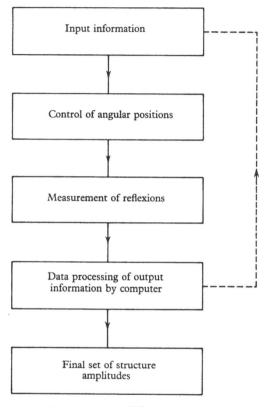

Fig. 2. Flow of operations in an automatic diffractometer system. For 'on-line' operation there is feedback from the processed data to the input via the computer, as indicated by the broken line.

conditions until these criteria are satisfied. The data are then reduced to a set of structure amplitudes as in the off-line system. The on-line system is indicated by the broken line in Fig. 2.

Just as there are two different classes of computers, analogue and digital, so is there a similar choice of analogue or digital instruments in diffractometry. In analogue instruments the relationships between the setting angles are reconstructed by means of mechanical linkages, and the correct setting angles are generated by simple movements of the linkages: there is no need to compute the magnitudes of the setting angles, and the input information consists of the lattice parameters only, apart from various instrumental constants. The type of linkage governs the mode of operation, such as the sequence in which the reflexions are measured. A different mode requires changes in the mechanical design, and for this reason the analogue method of control is less flexible than digital control.

In the digital method of control, angle encoders convert angular positions of the diffractometer shafts into digital form. The setting of the shafts and the measuring of the reflexions are programmed according to previously prepared input instructions. These input instructions are usually supplied by a computer, as the formulae for the setting angles are too cumbersome for manual computation of any but a very small number of reflexions. Thus, without ready access to a computer, it may be preferable to use the analogue method of setting.

As regards the design of diffractometers, automatic or otherwise, there is a further choice concerning the type of diffraction geometry used. In the 'normal-beam equatorial' geometry (Fig. 3) the incident beam lies in the same plane (the equatorial plane) as that in which the detector moves, and one of the axes about which the crystal rotates (the ω-axis) is normal to the equatorial plane. This ω-axis is used as the oscillation axis in scanning across the reflexion. The crystal orientation is determined by the three Eulerian angles— ϕ, χ, ω—and these angles must be reset between measurements of separate reflexions. This geometry is particularly useful in neutron diffraction, where the motion of the detector, with its bulky fast-neutron shielding, is best restricted to a single, horizontal plane. In the 'inclination' (Fig. 4) type of geometry both the incident and diffracted beams are inclined at variable angles, $90° - \mu$ and $90° - \nu$ respectively, to the crystal oscillation axis, which, in contrast with the equatorial geometry, coincides with the goniometer-head axis ϕ. The crystal is rotated about the single axis ϕ to bring each reflecting plane to the measuring position and the detector is rotated about two independent axes, ν and Υ, to receive the diffracted beam. With inclination instruments used in one of the special settings (equi-inclination, normal-beam, flat-cone) discussed in Chapter 2, it is only necessary to alter one detector axis and one crystal axis in order to measure all the reflexions in a given reciprocal lattice level; for automatic operation it is usual to control these two axes automatically and to adjust the remaining axis by hand between the measurement of different levels.

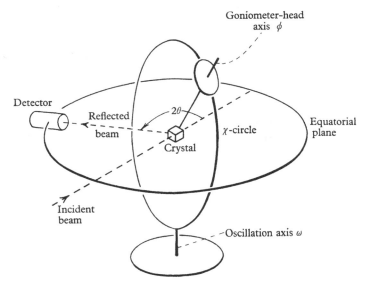

Fig. 3. Normal-beam equatorial geometry. The crystal is mounted on a gonio-meter-head attached to the ϕ-axis; the ϕ-circle moves round the vertical χ-circle, and the ϕ–χ assembly rotates as a whole about the vertical ω-axis. The detector moves in the horizontal, equatorial plane and the incident beam is normal to the crystal oscillation axis ω.

Once the crystal and detector are set correctly to record a given reflexion, one of three measuring procedures may be initiated to determine the magnitude of the integrated intensity. In the first (*stationary-crystal-stationary-detector* method) the peak intensity and the background in the vicinity of the reflexion are measured with the crystal and detector both stationary. The difference between the peak intensity and the background intensity is proportional to the integrated intensity, provided the incident beam has a wide and uniform angular distribution of intensity at the

specimen. These two conditions are difficult to satisfy simultaneously, and so, in spite of its speed and simplicity, the stationary-
crystal-stationary-detector method is rarely used. In the other two
measuring procedures, the crystal moves slowly through the
reflecting position as the reflected beam is recorded: the crystal
alone moves in the *moving-crystal-stationary-detector* method; the

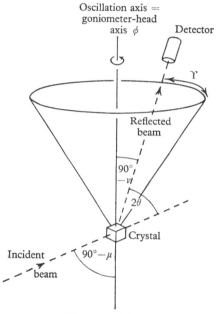

Fig. 4. Inclination geometry. The crystal is mounted on a goniometer-head
attached to a single shaft ϕ. The detector moves around a cone with variable
semi-angle, $90° - \nu$, and the incident beam likewise makes a variable angle,
$90° - \mu$, with the crystal oscillation axis ϕ.

detector moves at twice the angular velocity of the crystal in the
moving-crystal-moving-detector method. In these two scans a
complete profile of the reflexion is measured and the area of this
profile above the background is the required integrated intensity.

1.4. Accuracy of counter methods

One of the objects of crystal-structure determinations is often
the derivation of bond lengths between neighbouring atoms. To
measure bond lengths to an accuracy of 0·01 Å requires extremely

good experimental data, and for the study of thermal vibration amplitudes or electron density distributions even higher quality may be necessary. This search for accuracy has contributed to the swing back to counter methods of detection in X-ray crystallo-graphy. Modern scintillation and proportional counters are far superior to the earlier ionization chambers: it is now possible to count the individual X-ray quanta received by the detector and to record directly the intensity of each diffracted beam in terms of the number of quanta received per second. In the photographic method a direct determination of intensity is not possible. The blackness of a spot on a photographic film must be related to the intensity of the beam producing it, and, even in the most careful work, this intro-duces errors of perhaps 10 per cent in the estimated intensities. The photographic determination of X-ray intensities is fully described in the book *Crystal-Structure Analysis* by M. J. Buerger (1960).

The production of X-rays in an X-ray tube, or of thermal neutrons in a nuclear reactor, is a statistical process and is subject to random statistical fluctuations. If the detector records a total of N counts in a given time, the standard deviation of repeated measurements of this number is $N^{\frac{1}{2}}$, and the fractional standard deviation is $N^{-\frac{1}{2}}$. This means that at least 10,000 counts must be recorded to achieve a percentage standard deviation not exceeding 1 per cent.

The final accuracy of the experimental data obtained with a diffractometer is very often limited by systematic errors rather than by the random statistical error. Systematic errors arising from such factors as absorption, extinction, simultaneous reflexions or thermal diffuse scattering (which peaks at the same position as the Bragg reflexion and causes errors in estimating the background level under the reflexion) are difficult to calculate or to correct for experi-mentally. It may be possible to correct some of the systematic errors by repeating or extending the measurements under different experimental conditions (for instance, by using a different wave-length or by changing the azimuthal orientation of the reflecting planes), but very careful and painstaking work is necessary to reduce the relative errors of the structure factors to the region of 1 per cent. In later chapters we shall discuss the questions of random and systematic errors, and analyse the results of intensity measurements carried out on standard crystals with known structures.

1.5. Speed of counter methods

Nearly all X-ray diffractometers measure the diffraction peaks in succession, whereas photographic methods are capable of recording a complete reciprocal lattice level on the same film. On the other hand, diffractometers can be adapted readily to automatic methods of control: automatic diffractometers are capable of high speed and accuracy (although these two features tend to be mutually exclusive), and they eliminate much of the tedium and labour demanded by photographic work.

It is probably in the study of large biological molecules that the high speed of diffractometer methods is of greatest importance. The determination of the structure of even a relatively small protein requires the measurement of perhaps 250,000 reflexions: a data collection task of this magnitude could not readily be undertaken without an automatic diffractometer. An example from this field (H. C. Watson, private communication, 1965) illustrates the advantages of diffractometry in a particularly striking way.

The protein, glyceraldehyde 3-phosphate dehydrogenase, has a molecular weight of 140,000 and four molecules are contained in the unit cell of approximate dimensions $a = 150$ Å, $b = 140$ Å and $c = 80$ Å. There are approximately 3,500 independent reflexions from planes with spacings greater than 6·5 Å. These reflexions were measured first using two Buerger precession cameras operating simultaneously on two crystals. The resulting films were measured on a recording microdensitometer, the chart records analysed by experienced assistants, and the intensities punched on cards. The scaling of the measurements on different films and on different crystals, and the correction for Lorentz and polarization factors, were carried out by computer. (The data were not corrected for absorption.) Three months of full-time work by one scientist and one assistant were required to produce a list of 3,500 structure factors: this corresponds to an average output of about 50 reflexions/day. A total of 36 different crystals were used in this investigation since each crystal could withstand only 72 h in the X-ray beam.

The same reflexions were then measured on one crystal using a paper-tape-controlled three-circle diffractometer. The time taken—by one scientist—was 3 days, and the data were then already on paper tape, ready for further processing. Because a distant computer was used, an additional day was needed for this processing. This data collection rate of 1,000 reflexions/day can be enhanced by factors of three or five when employing multiple counter techniques (Phillips, 1964; Arndt, North & Phillips, 1964), without any increase in X-ray exposure to the specimen.

1.6. Cost of counter methods

In spite of the high capital costs of automatic-diffractometer installations, their use for data collection may give rise to real economic advantages. This is shown by analysing the data-collection problem discussed above.

(a) Photographic data collection (X-rays)

Capital cost of X-ray tube, two precession cameras, densitometer, darkroom equipment	£3,500
Annual costs	
Interest on capital at 5 % p.a.	£175
Depreciation (assuming writing-off the equipment in 10 years)	£350
Materials, spares, servicing	£200
Scientist's salary	£2,000
Full-time technician	£800
	£3,525 p.a.

Number of reflexions measured in one year at 3,500 reflexions in 3 months = 14,000
Hence cost per 1,000 reflexions = £250

(b) Diffractometer data collection (X-rays)

Capital cost of diffractometer, X-ray tube, electronic test equipment	£20,000
Annual costs	
Interest on capital at 5 % p.a.	£1,000
Depreciation (assuming writing-off the equipment in 7½ years)	£2,667
Materials, spares, servicing	£800
Scientist's salary	£2,000
Part-time electronics technician	£700
Part-time assistant	£500
	£7,667 p.a.

Number of reflexions measured per year at 5,000 per
week for 48 weeks = 240,000
Hence cost per 1,000 reflexions = £32

It must be emphasized that economies of this magnitude are
realized only if the diffractometer is fully employed: this in turn
implies that the laboratory in which it is installed has data-collection problems of an adequate magnitude, and that the instrument
is sufficiently reliable to function for a high fraction of the working
year. These two conditions as yet apply simultaneously in very few
laboratories.

Diffractometer data collection (neutrons)

The cost of providing a monochromatic neutron beam varies
from reactor to reactor and is very difficult to determine: however,
for a 20 MW reactor a figure of £10,000–£15,000 p.a. per diffractometer is perhaps appropriate. The costs of the remaining items in
the diffractometer installation are essentially the same as for X-rays,
and so we can estimate a total annual cost of about £20,000 per
diffractometer. Assuming that 500 reflexions are measured in one
week, the net cost per 1,000 reflexions is £1,000. This is a very
crude overall figure but it does serve to emphasize that neutrons
must only be used for those crystallographic problems—for instance,
the determination of magnetic structures, and the location of light
atoms in compounds containing heavy elements—which could not
be tackled with X-rays.

CHAPTER 2

DIFFRACTION GEOMETRY

We have referred already (see p. 7) to the two principal methods of measuring a set of Bragg reflexions. The first is the *inclination* method, which is related to the photographic Weissenberg technique. The second, the *normal-beam equatorial* method, has no counterpart in photographic work. In this chapter we shall describe the diffraction geometry associated with these two methods and derive formulae for the setting angles of the crystal and detector, both in a general form and in various simplified forms for special settings and particular crystal symmetries. We shall then compare the two geometries and show how the special settings which are used in either method are related to one another. Finally, we shall discuss the problem of measuring several reflexions at the same time with a diffractometer.

It is necessary to describe first the geometrical requirements for setting the crystal and detector and for measuring the reflexion. Bragg's law imposes certain geometrical conditions on the positions of the crystal and the detector, and these conditions must be satisfied before the measurement begins. Once the crystal and detector are correctly set, the reflexion is measured by counting the number of diffracted X-ray quanta or slow neutrons received by the detector as the crystal rotates uniformly through the Bragg reflecting region. These geometrical considerations are best described in terms of the reciprocal lattice and the Ewald sphere of reflexion.

2.1. General considerations

The reciprocal lattice

The unit cell of the direct lattice is the parallelepiped with edges $\mathbf{a}$, $\mathbf{b}$, $\mathbf{c}$. The magnitudes of these vectors are a, b, c and their interaxial angles are α, β, γ, where α is the angle between $\mathbf{b}$ and $\mathbf{c}$, β that between $\mathbf{c}$ and $\mathbf{a}$ and γ that between $\mathbf{a}$ and $\mathbf{b}$.

The vectors **a***, **b***, **c*** defining the unit cell of the reciprocal lattice are derived from **a**, **b**, **c** in the following way. The vector **a*** is normal to the plane containing **b** and **c**, and its magnitude **a*** is proportional to the reciprocal of the spacing of the (100) family of planes in the direct lattice. Thus

$$\mathbf{a^*} \cdot \mathbf{b} = \mathbf{a^*} \cdot \mathbf{c} = 0,$$

where the dot indicates a scalar product, and

$$|\mathbf{a^*}| = a^* = K/d_{100},$$

where K is a constant. Unit cells in the direct and reciprocal lattices are shown in Fig. 5.

In considering the reciprocal lattice in relation to the diffraction process, it is convenient to choose $K = \lambda$, the wavelength of the incident radiation. This choice will be used throughout this book. We have then

$$a^* = \lambda/d_{100}.$$

Similarly, **b*** is defined as a vector normal to the **ca**-plane and of magnitude

$$b^* = \lambda/d_{010},$$

and **c*** as normal to the **ab**-plane and of magnitude

$$c^* = \lambda/d_{001}.$$

The relations between the cell edges in the direct lattice and those in the reciprocal lattice are:

$$\mathbf{a^*} \cdot \mathbf{a} = \lambda, \quad \mathbf{a^*} \cdot \mathbf{b} = 0, \quad \mathbf{a^*} \cdot \mathbf{c} = 0,$$
$$\mathbf{b^*} \cdot \mathbf{a} = 0, \quad \mathbf{b^*} \cdot \mathbf{b} = \lambda, \quad \mathbf{b^*} \cdot \mathbf{c} = 0,$$
$$\mathbf{c^*} \cdot \mathbf{a} = 0, \quad \mathbf{c^*} \cdot \mathbf{b} = 0, \quad \mathbf{c^*} \cdot \mathbf{c} = \lambda,$$

and the explicit expressions for **a***, **b***, **c*** in terms of **a**, **b**, **c** are

$$\left. \begin{aligned} \mathbf{a^*} &= \lambda(\mathbf{b} \wedge \mathbf{c}/V), \\ \mathbf{b^*} &= \lambda(\mathbf{c} \wedge \mathbf{a}/V), \\ \mathbf{c^*} &= \lambda(\mathbf{a} \wedge \mathbf{b}/V). \end{aligned} \right\} \tag{2.1}$$

Here **b**∧**c** denotes the vector product of **b** and **c** and V is the volume of the unit cell in the direct lattice, that is, $V = \mathbf{a} \cdot \mathbf{b} \wedge \mathbf{c}$.

The reciprocal lattice has two properties which are particularly useful in considering diffraction problems. These properties,

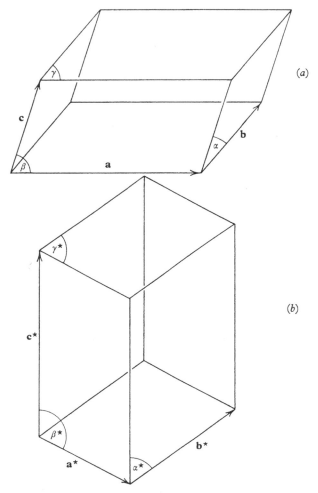

Fig. 5. Unit cells in (a) the direct lattice, (b) the reciprocal lattice. Right-handed system of axes.

which are proved in a number of standard textbooks (see, for example, James, 1962), are:

(i) The vector $\mathbf{d}^* = h\mathbf{a}^* + k\mathbf{b}^* + l\mathbf{c}^*$, which joins the origin of the reciprocal lattice to the point with co-ordinates hkl, is perpendicular to the family of planes in the direct lattice with indices (hkl).

(ii) The magnitude of $\mathbf{d}^*$ is λ/d, where d is the spacing of the (hkl) planes of the direct lattice.

If the reciprocal lattice co-ordinates hkl have a common factor n, the corresponding (hkl) planes in the direct lattice are understood, in the usual sense, as those with a spacing of $1/n$ times the spacing of the fundamental set with Miller indices h/n, k/n, l/n.

It is sometimes convenient to replace the vectors $\mathbf{a}^*$, $\mathbf{b}^*$, $\mathbf{c}^*$ by expressions involving their magnitudes a^*, b^*, c^* and their interaxial angles α^*, β^*, γ^*, where α^* is the angle between $\mathbf{b}^*$ and $\mathbf{c}^*$, β^* the angle between $\mathbf{c}^*$ and $\mathbf{a}^*$, and γ^* the angle between $\mathbf{a}^*$ and $\mathbf{b}^*$. Thus the equation

$$|\mathbf{d}^*| = |h\mathbf{a}^*+k\mathbf{b}^*+l\mathbf{c}^*| = \lambda/d \qquad (2.2)$$

can be rewritten as

$$d^{*2} = h^2a^{*2}+k^2b^{*2}+l^2c^{*2}+2hk\,a^*b^*\cos\gamma^*+2kl\,b^*c^*\cos\alpha^*$$
$$+2lh\,c^*a^*\cos\beta^* = \lambda^2/d_{hkl}^2. \qquad (2.3)$$

In general, we shall express formulae in terms of quantities in reciprocal space. Unit length in reciprocal space, which is dimensionless because of the choice $K = \lambda$, will be denoted 1 r.l.u. (reciprocal lattice unit). To convert formulae involving a^*, b^*, c^*, α^*, β^*, γ^* into corresponding formulae involving the quantities a, b, c, α, β, γ of the direct lattice, the equations (2.4) and (2.5) are employed:

$$\left.\begin{aligned} a^* &= \frac{\lambda bc\sin\alpha}{V}, \\[2mm] b^* &= \frac{\lambda ca\sin\beta}{V}, \\[2mm] c^* &= \frac{\lambda ab\sin\gamma}{V}, \end{aligned}\right\} \qquad (2.4)$$

where V = volume of direct cell
$$= abc\{1 + 2\cos\alpha\cos\beta\cos\gamma - \cos^2\alpha - \cos^2\beta - \cos^2\gamma\}^{\frac{1}{2}},$$

and
$$\left.\begin{aligned} \cos\alpha^* &= \frac{\cos\beta\cos\gamma - \cos\alpha}{\sin\beta\sin\gamma}, \\[2mm] \cos\beta^* &= \frac{\cos\gamma\cos\alpha - \cos\beta}{\sin\gamma\sin\alpha}, \\[2mm] \cos\gamma^* &= \frac{\cos\alpha\cos\beta - \cos\gamma}{\sin\alpha\sin\beta}. \end{aligned}\right\} \qquad (2.5)$$

The same equations hold when the starred and unstarred quantities are interchanged.

Geometrical conditions for observing Bragg reflexion. The Ewald sphere

Bragg's law

$$\lambda = 2d\sin\theta, \tag{2.6}$$

expresses the condition that radiation of wavelength λ is diffracted at glancing angle θ by the planes of the direct lattice of spacing d. The law can be interpreted geometrically in terms of the reciprocal lattice and the Ewald sphere of reflexion. This interpretation is particularly convenient in calculating the setting angles of the crystal and detector for observing the Bragg reflexion, and it will be used extensively throughout this book.

The Ewald sphere is defined with reference to the unit vector $\mathbf{s_0}$, which lies in the direction of the incident beam. If $\mathbf{s_0}$ terminates at the origin O of the reciprocal lattice and has its starting point at C, then the Ewald sphere is the sphere with centre C and radius 1 r.l.u.

Suppose that the reciprocal lattice point P with co-ordinates hkl lies on the surface of the Ewald sphere (Fig. 6). OP is normal to the family of planes (hkl) in the direct lattice and so the angle between the incident beam and OP is $90° - \theta'$, where θ' is the glancing angle of incidence. The length of the diameter OA is 2 r.l.u. and of OP is λ/d, so that in the right-angled triangle APO

$$\cos(90° - \theta') = \lambda/2d. \tag{2.7}$$

Comparison of (2.6) and (2.7) shows that the glancing angle θ' is equivalent to the Bragg angle θ. This is only true if P lies on the surface of the sphere.

Thus the condition for the family of planes (hkl) to diffract radiation of wavelength λ is equivalent to the requirement that the hkl point of the reciprocal lattice lies on the surface of the Ewald sphere of reflexion. We can represent the crystal by a collection of reciprocal lattice points, each point corresponding to a different family of planes in the direct lattice. As the reciprocal lattice rotates it sweeps through the Ewald sphere, and a reflexion occurs each time a reciprocal lattice point cuts the sphere.

In the idealized situation described above the reciprocal lattice points are geometrical points. In practice, the incident beam contains a range of directions and wavelengths and the crystal is of finite extent with reflecting planes that are not exactly parallel: these facts can be accommodated in our geometrical picture by regarding the reciprocal lattice 'points' as small regions of finite size in reciprocal space. Radiation is then diffracted when any part of the small regions lie on the surface of the sphere.

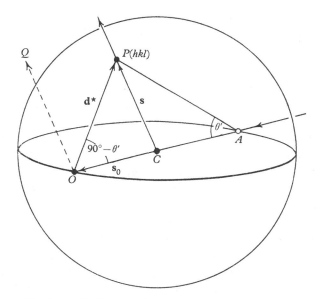

Fig. 6. Ewald sphere of reflexion. O is the origin of reciprocal space, s_0 and s are unit vectors along the incident and reflected beams, and P is a point with co-ordinates hkl with respect to the axes of the reciprocal lattice.

The diffracted beam lies along a direction which is at an angle 2θ to the incident beam and lies in a plane containing the incident beam and the reflecting normal. Fig. 6 shows that this direction is along the line CP, joining the centre of the Ewald sphere to the hkl reciprocal lattice point. To receive the diffracted radiation the detector must be placed along CP, where C is taken as the position of the crystal. In discussing angular relationships with the aid of the Ewald construction, we are at liberty to draw the diffracted beam as originating either at C, the crystal position, or at O, the origin of the reciprocal lattice. The former practice is generally more convenient, but there are cases where it is preferable to draw the diffracted beam along OQ, parallel to CP (see Fig. 6).

If $\mathbf{s}$ is a unit vector in the direction of the diffracted beam, the relation

$$\mathbf{d^*} = \mathbf{s} - \mathbf{s_0} \qquad (2.8)$$

holds at the reflecting position for the (hkl) plane, where

$$\mathbf{d^*} = h\mathbf{a^*} + k\mathbf{b^*} + l\mathbf{c^*}.$$

The 'scattering vector' $\mathbf{S}$ is defined as $\mathbf{s} - \mathbf{s_0}$, so that the Bragg condition (2.6) is equivalent to bringing the vectors $\mathbf{S}$ and $\mathbf{d^*}$ into coincidence.

Geometrical conditions for measuring Bragg reflexion

The integrated intensity of a reflexion, also known as the integrated reflexion, is proportional to the total energy reflected by the crystal as it passes with uniform angular velocity through the Bragg reflecting position. (A more complete definition is given on p. 234.) The integrated intensity is measured by recording the number of quanta entering the detector as the crystal rotates through a small angular range 2Δ about the Bragg position. During this movement of the crystal the detector can be either kept stationary or given a small movement related to that of the crystal. Accordingly, there are two principal moving-crystal measuring procedures. In the *moving-crystal-stationary-detector* procedure (the *ω-scan* in the normal-beam equatorial geometry) the detector remains fixed at an angle 2θ to the incident beam during the rotation of the crystal. In the alternative *moving-crystal-moving-detector* procedure (the *ω/2θ-scan* or *2θ-scan* in the normal-beam equatorial geometry) the detector shaft is coupled to the crystal shaft by a 2:1 linkage or gear train, so that the crystal rotates from $\theta - \Delta$ to $\theta + \Delta$ while the detector moves from $2\theta - 2\Delta$ to $2\theta + 2\Delta$. The detector window widths, while different in these two scans, must be sufficient in both cases to accept the full angular spread of the diffracted beam.

The crystal rotation through 2Δ can take place about any axis not coincident with the normal to the reflecting plane. In the normal-beam equatorial method the crystal rotates about an axis lying in the reflecting plane, and the velocity with which the reciprocal lattice point passes through the Ewald sphere is a function of θ only. A valuable feature of this method is that the view of the source is the same for all reflexions (Lang, 1954). In the general inclination method, on the other hand, the crystal rotates about the goniometer-head axis, which is inclined at a variable angle to the reflecting plane, and the magnitude of this angle depends on the particular reflexion and on the special inclination setting adopted for measuring it. The Lorentz factor, expressing the time spent by the reciprocal lattice point in passing through the Ewald sphere, has a more complicated form than in the normal-beam equatorial method (see p. 278). Moreover, reciprocal lattice points lying on or very close to the goniometer-head axis cannot cut across the Ewald sphere, and so these points are not accessible to measurement without re-orientating the crystal.

A third procedure for measuring the integrated intensity is to keep the crystal stationary at the Bragg position and to use a strongly convergent incident beam: the recorded intensity in this *stationary-crystal-stationary-detector* technique is then equivalent to that for a parallel beam and oscillating crystal (see Fig. 7). In X-ray work the effective source is a foreshortened view of a line focus, and the convergent beam can be produced merely by viewing the focus with a wide take-off angle.

The stationary-crystal, stationary-detector technique requires a uniform emission of X-rays from a substantial length of the line focus. This condition is difficult to meet in practice, and X-ray measurements made with an oscillating crystal technique tend to be better than those obtained with the stationary-crystal procedure. In neutron diffraction, where the incident beam is reflected by a plane-crystal monochromator, the divergence of the beam is considerably less than that required to give flat-topped diffraction peaks; the divergence can be increased by bending the monochromator but this is likely to give a non-uniform angular distribution of neutron flux. For these reasons the stationary-crystal technique is rarely used in either X-ray or neutron diffraction, in spite of the greater speed and simplicity it offers in data collection.

In all three measuring procedures, in addition to obtaining the diffracted intensity, we must measure the background intensity (see Chapter 8) in the immediate neighbourhood of the Bragg position. The difference between the intensities of the peak and of the background then gives the true integrated intensity.

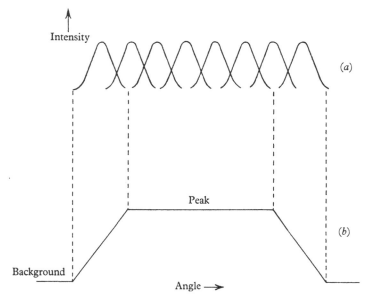

Fig. 7. Principle of stationary-crystal, stationary-detector procedure. The individual profiles of the reflexion in (a), each contributed by a different portion of the convergent incident beam, are summed in (b) to give a flat-topped peak. The integrated intensity is proportional to the peak level minus the background level in (b) (after Buerger, 1960).

2.2. Cylindrical polar co-ordinates in reciprocal space

We can now consider the problem of applying the Ewald construction to the determination of the setting angles for the crystal and detector. General formulae for these setting angles are derived in §§2.3 and 2.4, but to apply these formulae we need first to convert the co-ordinates hkl in reciprocal space to cylindrical polar co-ordinates, taking the 'goniometer-head axis' as the polar axis of cylindrical co-ordinates. This is the axis of the diffractometer on which the goniometer-head carrying the crystal is directly mounted, and we shall always denote the axis by the symbol ϕ.

The cylindrical co-ordinates are ξ, ζ, τ, where ξ is the radial co-ordinate, ζ the axial co-ordinate and τ the angular co-ordinate (see Fig. 8). These symbols are used by Waser (1951) and others; in chapter 4 of the *International Tables*, volume II, the symbol ϕ is used instead of τ. For our purpose, it is more convenient to adopt Waser's notation and to reserve ϕ for rotation about the

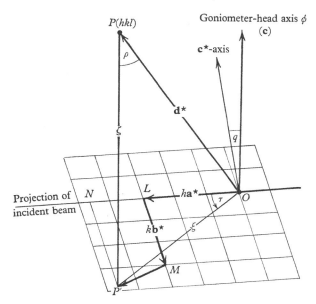

Fig. 8. View of reciprocal-lattice point *hkl* above **a*b*** plane.

goniometer-head axis in both the inclination and the normal-beam equatorial methods. The angle τ is measured in a clockwise direction, looking along the positive direction of the polar axis, with zero τ coinciding with the projection of the incident beam on the plane normal to the polar axis.

We shall assume that the goniometer-head axis ϕ is parallel to the *c*-axis of the crystal and that for $\tau = 0$ the **a***-axis is along the trace of the incident beam. Thus the ϕ-axis is normal to the plane **a*b*** of the reciprocal lattice. In using the inclination geometry to measure a set of Bragg reflexions it is customary to aline the crystal with the ϕ-axis normal to a reciprocal-lattice plane: this type of alinement is not necessary with the normal-beam equatorial geometry and, in fact, it may be preferable to avoid it and to place the crystal at an arbitrary orientation on the goniometer head.

However, in comparing the two kinds of geometry it is convenient to consider an identical orientation of the crystal in the two geometries. In a later section (p. 51) we discuss the general case in which the crystal assumes an arbitrary orientation in the equatorial geometry.

The expression for $\xi\zeta\tau$ in terms of hkl and the lattice parameters of a triclinic crystal can be derived from Figs. 8 and 9. Fig. 8 is a perspective view looking onto the $\mathbf{a^*b^*}$-plane. Fig. $9a$ is a projection on the $\mathbf{a^*b^*}$-plane and Fig. $9b$ is a stereogram giving the angular relationships between p and q, where p is the angle between $\mathbf{a^*}$ and the projection of $\mathbf{c^*}$ on the $\mathbf{a^*b^*}$-plane and q is the angle between $\mathbf{c}$ and $\mathbf{c^*}$. From the Napierian triangle ABC in Fig. $9b$, bounded by $90° - q$, p and β^*:

$$\sin q \cos p = \cos \beta^*, \tag{2.9}$$

$$\sin q \sin p = -\cos \alpha \sin \beta^*, \tag{2.10}$$

and
$$\cos q = \sin \alpha \sin \beta^*. \tag{2.11}$$

The ζ co-ordinate is the magnitude of the vector $l\mathbf{c^*}$, projected on the goniometer-head axis. Thus

$$\zeta = lc^* \cos q,$$

$$= lc^* \sin \alpha \sin \beta^*, \tag{2.12}$$

from equation (2.11). ζ can be positive or negative, depending on the sign of l.

The ξ co-ordinate can now be derived from the relation

$$|\mathbf{d^*}|^2 = \xi^2 + \zeta^2.$$

Substituting for $|\mathbf{d^*}|^2$ from equation (2.3) and for ζ^2 from (2.12) gives

$$\xi = [h^2 a^{*2} + k^2 b^{*2} + l^2 c^{*2} (1 - \sin^2 \alpha \sin^2 \beta^*) + 2hka^* b^* \cos \gamma^*$$
$$+ 2klb^* c^* \cos \alpha^* + 2lhc^* a^* \cos \beta^*]^{\frac{1}{2}}. \tag{2.13}$$

ξ must be positive, and so the positive root is taken in (2.13).

The third co-ordinate τ follows from Fig. $9a$ using the relation

$$\tan \tau = \frac{P'N}{ON} = \frac{kb^* \sin \gamma^* + lc^* \sin q \sin p}{ha^* + kb^* \cos \gamma^* + lc^* \sin q \cos p}.$$

Substituting for $\sin q \cos p$ from (2.9) and for $\sin q \sin p$ from (2.10) gives:

$$\tan\tau = \frac{\sin\tau}{\cos\tau} = \frac{kb^*\sin\gamma^* - lc^*\cos\alpha\sin\beta^*}{ha^* + kb^*\cos\gamma^* + lc^*\cos\beta^*}. \qquad (2.14)$$

τ lies in the range 0–360°. The correct quadrant follows by giving $\sin\tau$ the sign of the numerator in (2.14) and $\cos\tau$ the sign of the denominator.

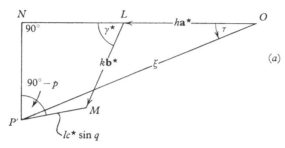

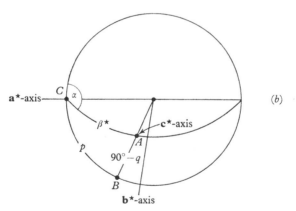

Fig. 9. (a) Vectors in **a*b***-plane; (b) stereogram projected on to **a*b***-plane.

Equations (2.13), (2.12) and (2.14) are the required expressions for the cylindrical co-ordinates ξ, ζ, τ in terms of *hkl* and the lattice parameters. To obtain formulae related to the parameters of the reciprocal cell only, the angle α is replaced by

$$\cos^{-1}\left(\frac{\cos\beta^*\cos\gamma^* - \cos\alpha^*}{\sin\beta^*\sin\gamma^*}\right).$$

The resultant expressions are cumbersome, but considerable simplification occurs for crystal systems of higher symmetry than triclinic.

Table I gives expressions for ξ, ζ, τ for the different crystal systems. The table includes expressions for both the first and second settings of the monoclinic system. In both settings the unique diad axis is along the goniometer-head axis, but is labelled **c** in the first setting (with τ measured from the **a*** axis as zero) and **b** in the second setting (with zero τ along **c***). A similar table has been constructed by Prewitt (1960).

2.3. Inclination method

This is the counter diffractometer version of the photographic Weissenberg method, which is very fully described by Buerger (1942). The Bragg reflexions are measured by rotating the crystal with uniform angular velocity about the goniometer-head axis, ϕ. The incident beam is inclined at an angle $90° - \mu$ to the ϕ-axis and the Bragg reflexion occurs in a direction at $90° - \nu$ to this axis.

If the crystal is mounted with a zone axis, say [001] or the c-axis, along the goniometer-head axis, the reciprocal lattice layers or levels at $l = 0, 1, 2, \ldots$ are normal to the ϕ-axis. As the crystal rotates, these levels intersect the Ewald sphere in circles ('reflecting circles') and give reflexions lying in cones of semi-angles $90° - \nu$, where each level is associated with a particular value of ν. To pick up a reflexion in a given level the detector is set at the correct angle $90° - \nu$ to the ϕ-axis and is then moved through an angle Υ about an axis concentric with the ϕ-axis. By measuring the reflexions level-by-level, μ and ν can be kept fixed and only two angles, ϕ for the crystal and Υ for the detector, need be varied within each level (see Fig. 4 on p. 9).

If the crystal is mounted in an arbitrary orientation on the goniometer head, so that the ϕ-axis does not coincide with a zone-axis, it is still possible to measure the reflexions, varying three angles ϕ, ν and Υ between measurements. μ can have a fixed value. However, this procedure would be inconvenient as it requires a variation of all three angles, even within a given level. We shall discuss only that orientation of the crystal in which a zone axis is along the ϕ-axis.

TABLE I. *Cylindrical co-ordinates of reciprocal lattice point (hkl) for various crystal systems*

Crystal system	ξ	ζ	τ
Triclinic; polar axis, **c**	$\left[h^2a^{*2}+k^2b^{*2}+\dfrac{l^2c^{*2}}{\sin^2\gamma^*}(\cos^2\alpha^* \right.$ $+\cos^2\beta^*-\cos^2\gamma^*$ $+2hka^*b^*\cos\gamma^*+2klb^*c^*\cos\alpha^*$ $\left. +2lhc^*a^*\cos\beta^*\right]^{\frac12}$	$\dfrac{lc^*}{\sin\gamma^*}[1-\cos^2\alpha$ $-\cos^2\beta^*-\cos^2\gamma^*$ $+2\cos\alpha^*\cos\beta^*\cos\gamma^*]^{\frac12}$	$\tan^{-1}\left\{\dfrac{kb^*\sin\gamma^*+lc^*\left(\dfrac{\cos\alpha^*-\cos\beta^*\cos\gamma^*}{\sin\gamma^*}\right)}{ha^*+kb^*\cos\gamma^*+lc^*\cos\beta^*}\right\}$
Monoclinic, first setting $(\alpha^*=\beta^*=90^\circ)$; polar axis, **c**	$[h^2a^{*2}+k^2b^{*2}+2hka^*b^*\cos\gamma^*]^{\frac12}$	lc^*	$\tan^{-1}\left[\dfrac{kb^*\sin\gamma^*}{ha^*+kb^*\cos\gamma^*}\right]$
Monoclinic, second setting $(\alpha^*=\gamma^*=90^\circ)$; polar axis, **b**	$[h^2a^{*2}+l^2c^{*2}+2lhc^*a^*\cos\beta^*]^{\frac12}$	kb^*	$\tan^{-1}\left[\dfrac{ha^*\sin\beta^*}{lc^*+ha^*\cos\beta^*}\right]$
Hexagonal $(a^*=b^*;$ $\alpha^*=\beta^*=90^\circ; \gamma^*=60^\circ)$; polar axis, **c**	$[h^2+k^2+hk]^{\frac12}a^*$	lc^*	$\tan^{-1}\left[\dfrac{\sqrt{3}k}{2h+k}\right]$
Orthorhombic $(\alpha^*=\beta^*=\gamma^*=90^\circ)$; polar axis, **c**	$[h^2a^{*2}+k^2b^{*2}]^{\frac12}$	lc^*	$\tan^{-1}\left[\dfrac{kb^*}{ha^*}\right]$
Tetragonal $(a^*=b^*;$ $\alpha^*=\beta^*=\gamma^*=90^\circ)$; polar axis, **c**	$[h^2+k^2]^{\frac12}a^*$	lc^*	$\tan^{-1}\left[\dfrac{k}{h}\right]$
Cubic $(a^*=b^*=c^*;$ $\alpha^*=\beta^*=\gamma^*=90^\circ)$; polar axis, **c**	$[h^2+k^2]^{\frac12}a^*$	lc^*	$\tan^{-1}\left[\dfrac{k}{h}\right]$

For monoclinic system, second setting, polar axis is along **b** and zero τ is along **c***.

For all other cases polar axis is along **c** and zero τ is along **a***.

ζ is always positive. ζ takes the same sign as the index l. τ lies between 0 and 360°: to obtain the correct quadrant the numerator in the expression for $\tan\tau$ gives the sign of $\sin\tau$ and the denominator the sign of $\cos\tau$.

General setting angles of crystal and detector

We wish to derive formulae giving the setting angles ϕ, ν, Υ in terms of the inclination angle μ and the cylindrical co-ordinates $\xi\zeta\tau$ of the *hkl* point. In the *general* inclination geometry μ can take an arbitrary value.

The reference orientation of the crystal is chosen with the **c**-axis along the goniometer-head axis, $+$**c** upwards in Fig. 4, and the **a***-axis along the projection of the incident beam in the plane normal to the goniometer-head axis. ϕ is the angle of rotation from this standard orientation, measured in a clockwise manner looking along the positive **c** direction. Υ is the angle which the detector moves round from the 'straight through' position: positive Υ corresponds to clockwise rotation of the detector arm looking along **c**, or counterclockwise rotation as seen looking down from above in Fig. 4. Thus Υ lies between 0 and 180° for instruments with counter-clockwise 2θ motion and between 0 and $-180°$ for instruments with clockwise 2θ motion. By defining Υ in this way we can derive formulae which apply to both left-handed and right-handed instruments.

Fig. 10a is a perspective drawing of the Ewald sphere, showing the reciprocal lattice point P ($=hkl$) at the standard orientation of the crystal and the same point S after rotation through ϕ to bring it to the surface of the sphere. S lies in the *l*-level of the reciprocal lattice, normal to the rotation axis. The figure also shows the zero level, passing through the origin O of the reciprocal lattice, and the equatorial level which is normal to the rotation axis and passes through the centre C of the Ewald sphere. The setting angles ν, Υ of the detector are given by

$$\sin\nu = \zeta + \sin\mu, \qquad (2.15)$$

and
$$\cos\Upsilon = \frac{CQ^2 + CM^2 - QM^2}{2CQ.CM}$$

$$= \frac{[1 - (\zeta + \sin\mu)^2] + \cos^2\mu - \xi^2}{2[1 - (\zeta + \sin\mu)^2]^{\frac{1}{2}}\cos\mu},$$

or
$$\cos\Upsilon = \frac{2\cos^2\mu - 2\zeta\sin\mu - \xi^2 - \zeta^2}{2\cos\mu(\cos^2\mu - 2\zeta\sin\mu - \zeta^2)^{\frac{1}{2}}}. \qquad (2.16)$$

The inclination angles μ, ν are restricted to the first quadrant, 0–90°, and the positive root is taken in the denominator of (2.16).

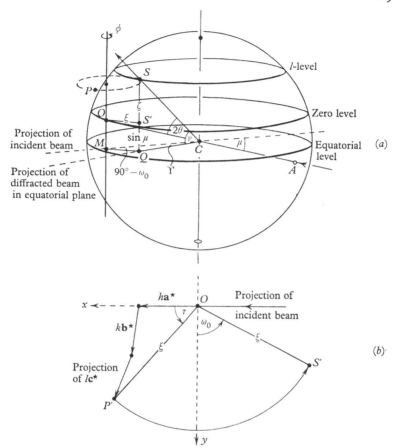

Fig. 10. (a) Ewald sphere with crystal in general inclination setting. The crystal is mounted with its **c**-axis along the goniometer-head axis ϕ. S is the reciprocal-lattice point hkl in the reflecting position, and P in the reference position. (b) Zero level in general inclination setting. P' is the projection of the hkl point in the reference position of the crystal and S' the projection of hkl in the reflecting position. ϕ is the angle between OP' and OS'.

The correct quadrant of Υ is determined by the sense of the 2θ motion, clockwise or anti-clockwise (see above).

The crystal setting angle ϕ is derived from Fig. 10b, which is a projection of Fig. 10a on the zero level. The points P and S project as P' and S'. The Cartesian axes xyz are laboratory axes with the z-axis parallel to **c** and the x-axis initially along **a***. ϕ is related to

the angle ω_0 between the y-axis and OS' in Fig. 10b by the equation
$$\phi = 180° - \tau - (90° - \omega_0). \tag{2.17}$$

From Fig. 10a

$$\cos(90° - \omega_0) = \frac{QM^2 + CM^2 - CQ^2}{2QM.CM}$$

$$= \frac{\xi^2 + \cos^2\mu - (1 - \zeta^2 - \sin^2\mu - 2\zeta\sin\mu)}{2\xi\cos\mu}$$

$$= \frac{\xi^2 + \zeta^2 + 2\zeta\sin\mu}{2\xi\cos\mu}. \tag{2.18}$$

Combining (2.17) and (2.18):

$$\phi = 180° - \tau - \cos^{-1}\left(\frac{\xi^2 + \zeta^2 + 2\zeta\sin\mu}{2\xi\cos\mu}\right). \tag{2.19}$$

The last term of (2.19) lies in the first quadrant for counter-clockwise 2θ motion and in the fourth quadrant for clockwise motion.

The setting angles for a crystal of any symmetry, examined in the general inclination setting, are readily derived by combining (2.15), (2.16) and (2.19) with the formulae for $\xi\zeta\tau$ in Table I.

We have derived the formulae for the setting angles without placing any restriction on the inclination angle μ. Thus μ can be assigned an arbitrary value before each hkl reflexion is measured. However, if we alter μ, the whole diffractometer assembly tilts with respect to the incident beam, and so it is mechanically desirable to carry out the intensity measurements at a minimum number of μ settings. Moreover, with μ fixed, the inclination angle ν of the detector is also fixed, in accordance with equation (2.15), for a given reciprocal lattice level which is normal to the goniometer-head axis ($\zeta = $ constant). For these reasons it is customary to assign μ a particular value for each level, measuring the hkl reflexions level by level and varying only the angles ϕ (for the crystal) and Υ (for the detector) within each level. The particular value chosen for μ gives rise to the normal-beam, equi-inclination, anti-equi-inclination and flat-cone settings.

Normal-beam setting ($\mu = 0$). The normal-beam setting is so-called, because the incident beam strikes the crystal at 90° to the axis of rotation. Putting $\mu = 0$ in (2.15), (2.16) and (2.19) gives the following expressions for the three setting angles:

$$
\left.
\begin{aligned}
\nu &= \sin^{-1}\zeta, \\
\Upsilon &= \cos^{-1}\left(\frac{2-\xi^2-\zeta^2}{2(1-\zeta^2)^{\frac{1}{2}}}\right), \\
\phi &= 180° - \tau - \cos^{-1}\left(\frac{\xi^2+\zeta^2}{2\xi}\right).
\end{aligned}
\right\}
\tag{2.20}
$$

For the zero level, $\zeta = 0$, the detector angle Υ is equal to 2θ. An instrument used in the normal-beam setting can be considered equally well as the diffractometer version of the photographic Weissenberg normal-beam method or the photographic rotating-crystal method.

A fundamental weakness of the normal-beam method is that the zero level only can be fully explored. A high proportion of the reflexions on upper levels lie in blind regions: they can be measured only by remounting the crystal in a new orientation. These blind regions are shown in Fig. 11. The small sphere in this figure represents the Ewald sphere of reflexion and the large sphere is the limiting sphere, of radius 2 r.l.u. and centre at the origin of the reciprocal lattice. The limiting sphere includes all *hkl* points up to the maximum Bragg angle of 90°. For $\zeta \neq 0$ there is an annular blind region around the rim of the reflecting circle at high Bragg angles, and an inner blind region at the centre of this circle of radius $1-(1-\zeta^2)^{\frac{1}{2}}$.

Equi-inclination setting $(\mu = -\nu)$. The incident and diffracted beams are equally inclined to the positive direction of the rotation axis. The setting angles in (2.15), (2.16) and (2.19) reduce to

$$
\left.
\begin{aligned}
\sin\nu &= \tfrac{1}{2}\zeta, \\
\Upsilon &= 2\sin^{-1}\left(\frac{\xi}{2\cos\mu}\right), \\
\phi &= 180° - \tau - \cos^{-1}\left(\frac{\xi}{2\cos\mu}\right).
\end{aligned}
\right\}
\tag{2.21}
$$

The angles Υ and ϕ are independent of ζ, so that there is a resemblance between the different levels, which in photographic work leads to simplifications in interpreting photographs of the various levels (Buerger, 1942).

There are no blind regions within the limiting sphere, because the axis of rotation ϕ passes through the reflecting circle of the l-level (Fig. 12). However, reciprocal lattice points close to the rotation axis have a large Lorentz factor and cannot be measured accurately. The equi-inclination setting is also liable to give intensity errors arising from simultaneous reflexions (see p. 251).

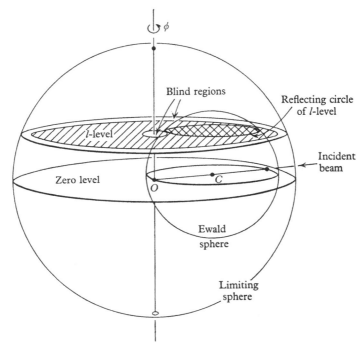

Fig. 11. Normal-beam setting, showing zero level and l-level. The l-level intersects the Ewald sphere in the cross-hatched region; as this region rotates the shaded area is scanned, leaving blind regions inside limiting sphere.

Such errors occur if at least two reciprocal lattice points lie simultaneously on the sphere of reflexion. This condition is automatically satisfied if the crystal is mounted in the equi-inclination setting with $\mathbf{c}^*$ (which coincides with $\mathbf{c}$ for orthogonal crystals) along the axis of rotation (Yakel & Fankuchen, 1962). In measuring the reflexions in the l-level the $00l$ point will then always lie on the surface of the Ewald sphere, and when hkl is in the reflecting position the *hk*o point is on the Ewald sphere too, as shown in Fig. 12. The simultaneous presence of *hkl, hk*o and oo*l* on the Ewald sphere can lead to serious errors in measuring *hkl* alone. A similar difficulty also occurs in the normal-beam setting: if **c*** is along the axis of rotation, *hkl* and *hkl̄* lie on the Ewald sphere simultaneously.

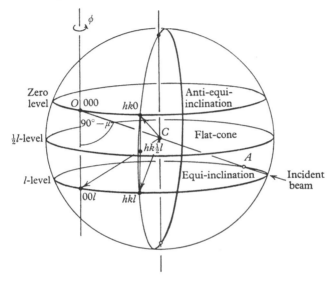

Fig. 12. Ewald sphere showing equi-inclination, flat-cone and anti-equi-inclination levels.

Anti-equi-inclination setting ($\mu = \nu$). From equation (2.15), $\zeta = 0$ if $\mu = \nu$. Consequently, this setting can only be used for reflexions in the zero level and is limited to the collection of two-dimensional data. The incident and diffracted beams are on opposite sides of the equatorial level, and are equally inclined to the rotation axis. The setting angles (equations 2.16 and 2.19) reduce to

$$
\left.
\begin{aligned}
\Upsilon &= 2\sin^{-1}\left(\frac{\xi}{2\cos\mu}\right), \\
\phi &= 180° - \tau - \cos^{-1}\left(\frac{\xi}{2\cos\mu}\right),
\end{aligned}
\right\}
\tag{2.22}
$$

which are the same expressions as for the equi-inclination setting.

Flat-cone setting ($\nu = 0$). If the incident beam makes an angle of $90° - \mu$ with the rotation axis, where μ is given by

$$\sin\mu = -\zeta, \tag{2.23}$$

then from equation (2.15) the inclination angle ν of the diffracted beam is zero. Thus for each level the diffracted beam is normal to the rotation axis and lies in a 'flat cone'. The l-level defined by (2.23) coincides with the plane passing through the centre of the Ewald sphere, normal to the goniometer-head axis, and the detector moves in this plane. The angular position of the detector is

$$\Upsilon = \cos^{-1}\left(\frac{2 - \xi^2 - \zeta^2}{2(1 - \zeta^2)^{\frac{1}{2}}}\right), \tag{2.24}$$

and the setting angle of the crystal is given by

$$\phi = 180° - \tau - \cos^{-1}\left(\frac{\xi^2 - \zeta^2}{2\xi\cos\mu}\right), \tag{2.25}$$

where $\sin\mu = -\zeta$.

Blind regions occur, as in the normal-beam setting, in measuring upper levels. The radius of the circular blind region at the centre of the l-level is $1 - (1 - \zeta^2)^{\frac{1}{2}}$, which is the same magnitude as for the normal-beam setting.

Fig. 12 is a perspective drawing of the sphere of reflexion, showing the equi-inclination, flat-cone and anti-equi-inclination settings. Whenever a general l-level is measured in the equi-inclination setting, the zero level will be in the anti-equi-inclination setting. This is a special case of the situation which arises whenever two levels are located symmetrically on either side of the flat-cone level, and gives rise to the possibility of measuring reflexions simultaneously in pairs (see §2.6).

The orientations of the incident and diffracted beams with respect to the rotation axis of the crystal in the normal-beam, equi-inclination and flat-cone settings are illustrated in Fig. 13; this figure is drawn for the measurement of the $l = 2$ level, and so does not include the anti-equi-inclination setting which can only be used for the zero level, $l = 0$.

Table II summarizes the formulae for the setting angles derived in this section. The first column gives the inclination angle μ of

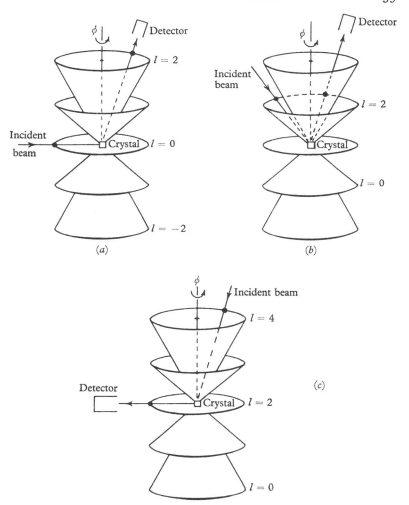

Fig. 13. Measurement of $l = 2$ level in (*a*) normal-beam, (*b*) equi-inclination, and (*c*) flat-cone settings. The diffraction cones for other levels are also shown.

the incident beam, which assumes a particular value for the normal-beam, equi-inclination and flat-cone settings. The remaining columns give the setting angles in terms of μ and the polar co-ordinates $\xi\zeta\tau$ of the *hkl* reciprocal lattice point. Combining Tables I and II leads to formulae for the setting angles in terms of *hkl* and the lattice parameters of the sample.

TABLE II. *Setting angles of crystal and detector: inclination geometry*

Setting	Inclination angle μ	Setting angle of crystal ϕ	Setting angles of detector	
			ν	Υ
General	Arbitrary	$180° - \tau - \cos^{-1}\left(\dfrac{\xi^2 + \zeta^2 + 2\zeta\sin\mu}{2\xi\cos\mu}\right)$	$\sin^{-1}(\zeta + \sin\mu)$	$\cos^{-1}\left(\dfrac{2\cos^2\mu - 2\zeta\sin\mu - \xi^2 - \zeta^2}{2\cos\mu(\cos^2\mu - 2\zeta\sin\mu - \zeta^2)^{\frac{1}{2}}}\right)$
Normal-beam	0	$180° - \tau - \cos^{-1}\left(\dfrac{\xi^2 + \zeta^2}{2\xi}\right)$	$\sin^{-1}\zeta$	$\cos^{-1}\left(\dfrac{2 - \xi^2 - \zeta^2}{2(1 - \zeta^2)^{\frac{1}{2}}}\right)$
Equi-inclination	$-\sin^{-1}(\tfrac{1}{2}\zeta)$	$180° - \tau - \cos^{-1}\left(\dfrac{\xi}{2\cos\mu}\right)$	$\sin^{-1}(\tfrac{1}{2}\zeta)$	$2\sin^{-1}\left(\dfrac{\xi}{2\cos\mu}\right)$
Anti-equi-inclination	Arbitrary	$180° - \tau - \cos^{-1}\left(\dfrac{\xi}{2\cos\mu}\right)$	Arbitrary ($=\mu$)	$2\sin^{-1}\left(\dfrac{\xi}{2\cos\mu}\right)$
Flat-cone	$\sin^{-1}(-\zeta)$	$180° - \tau - \cos^{-1}\left(\dfrac{\xi^2 - \zeta^2}{2\xi\cos\mu}\right)$	0	$\cos^{-1}\left(\dfrac{2 - \xi^2 - \zeta^2}{2(1 - \zeta^2)^{\frac{1}{2}}}\right)$

The angles μ, ν lie in the first quadrant.

Positive root is taken in expression for Υ. Υ lies in first and second quadrants for counter-clockwise 2θ motion and in third and fourth quadrants for clockwise 2θ motion.

Last term in expression for ϕ lies in first quadrant for counter-clockwise 2θ motion and in fourth quadrant for clockwise 2θ motion.

The **c**-axis of the crystal is along the goniometer-head axis, and at its reference position the crystal is orientated with the **a***-axis along the trace of the incident beam.

2.4. Normal-beam equatorial method

We now come to the second general class of diffraction geometry. Instruments using this kind of geometry are variously known as goniostats, single crystal orienters, three-circle or four-circle diffractometers. We shall use the term 'three-circle diffractometer' to denote an instrument in which the detector shaft is geared to one of the crystal shafts, and the term 'four-circle diffractometer' for an instrument with four independently driven shafts. During the measurement of each reflexion, carried out by oscillating the

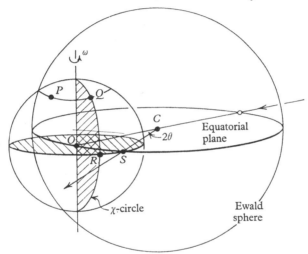

Fig. 14. The Ewald sphere, and the sphere through the reciprocal lattice point P with centre at the origin of the reciprocal lattice. In the symmetrical-A setting the ϕ rotation moves P to Q, the χ-rotation moves Q to R and the ω-rotation moves R to the reflecting position S.

crystal through the Bragg reflecting position, both the incident and diffracted beams are normal to the oscillation axis. Thus the incident and diffracted beams lie in the equatorial plane, which is the plane normal to the crystal oscillation axis and passing through the centre of the sphere of reflexion. We shall call this type of diffraction geometry Normal-Beam Equatorial Geometry (Fig. 3).

Fig. 14 shows the Ewald sphere of reflexion. The oscillation axis is denoted by ω: in contrast with the situation in the inclination method, this axis must be distinguished from the goniometer-head axis, which we denote again by ϕ. The detector rotates about the 2θ-axis, coincident with the ω-axis; its movement is restricted to the equatorial plane and a rotation through an angle 2θ from the incident beam brings the detector to the correct position for receiving the diffracted beam. To bring the crystal to the reflecting position for the *hkl* plane, the corresponding reciprocal lattice point P must move to S, where OS is in the equatorial plane and at $90° - \theta$ to the incident beam.

Conventions for setting angles

In a four-circle diffractometer the crystal has three rotational degrees of freedom. Three rotations are sufficient to give any vector, referred to axes in the crystal, any arbitrary orientation in laboratory space. The three circles (see Fig. 15) are the ω-circle, the χ-circle which is carried on the ω-circle and whose axis is normal to the ω-axis, and the ϕ-circle which is mounted on the χ-circle and carries the goniometer head supporting the crystal.

Before deriving expressions for the setting angles of the crystal we must adopt certain conventions for defining the setting angles and the standard orientation of the diffractometer with respect to the laboratory axes x, y, z. The x-axis is defined as the direction of the incident beam, outwards from the source, and is assumed to be horizontal. The z-axis is vertically upwards and the y-axis completes a right-handed system. ω is the angle of rotation of the χ-circle about a vertical diameter and χ is the angle made by the ϕ-axis with this diameter. We define the zero positions of ω and χ when the ϕ-axis is along z and below the crystal and the plane of the χ-circle is normal to x (Fig. 15). The zero value of ϕ, the angle of rotation about the goniometer-head axis, is arbitrary in the sense that it depends on the orientation of the crystal on the goniometer head. As in §2.3, we shall assume that the crystal is mounted with its **c**-axis along the goniometer-head axis and that $\phi = 0$ when the **a***-axis is along the positive x direction. The positive direction of the goniometer-head axis is defined as along $+z$ in Fig. 15.

We must also give conventions for the positive senses of rotation of the setting angles. When $\chi = 0$, ω and ϕ increasing represent clockwise rotation looking along the positive z-axis. When $\omega = 0$, χ increasing represents a clockwise rotation looking along the positive x-axis. These conventions are in accordance with the right-handed screw rule, previously used in defining the cylindrical co-ordinate τ and the inclination setting angles ϕ and Υ. The detector angle is measured as a clockwise rotation through 2θ about the positive z-axis: thus 2θ is positive for counter-clockwise movement of the detector as viewed from above the instrument, and negative for clockwise rotation. By allowing for both positive and negative values of θ we can use the formulae derived below for both left-handed and right-handed diffractometers.

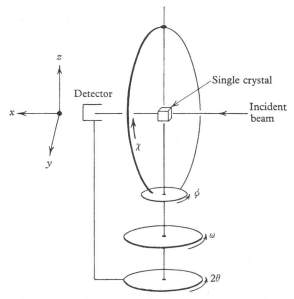

Fig. 15. Positive senses of rotation of ϕ, χ, ω, 2θ in equatorial method. The positive direction of the goniometer-head axis is along $+z$.

We note here that the orientation of the crystal, corresponding to the angles ω, χ, ϕ as defined above, is identical to that corresponding to $\omega + 180°$, $-\chi$, $\phi + 180°$.

Azimuthal orientation of reflecting plane

In the most commonly used setting, which we shall call the symmetrical-A setting or cone setting (Furnas & Harker, 1955), the χ-plane bisects the incident and diffracted beams at the measuring position. To bring the reciprocal lattice point P,

Fig. 14, to this position, the crystal is rotated through an angle ϕ so that P coincides with Q in the plane of the χ-circle, through an angle χ to move Q to R along the vertical χ-circle, and through an angle ω to move R to the surface of the sphere of reflexion at S. In general, however, it is not necessary that the hkl normal lies in the χ-plane at the measuring position. Within certain limits, to be determined below, the hkl normal can make any arbitrary angle ϵ with the χ-plane. This degree of freedom arises because the reflecting condition still holds if the crystal rotates about an axis normal to the reflecting plane; the choice of ϵ determines the azimuthal orientation ψ of the plane, measured as the angle of rotation about its normal. We shall take ϵ as positive when the χ-plane lies between the scattering vector and the incident beam (see Fig. 16a), so that the sense of increasing ϵ is the same as that for θ and ω.

To determine the permissible range of ϵ we refer to the stereogram in Fig. 16b, showing the crystal rotations ω, χ, ϕ which are necessary to bring the reciprocal lattice vector OP to the reflecting position OS lying along the scattering vector $\mathbf{S} = \mathbf{s} - \mathbf{s}_0$ in Fig. 16a. The ϕ-rotation moves the reciprocal lattice point P to Q, where Q lies at the point of intersection of two small circles, one through P and normal to the vertical z-axis and the other through R and normal to the x-axis. The χ-rotation moves Q to R, and the ω-rotation moves R to the reflecting position S at $90° - \theta$ to the incident beam. (Fig. 24 on p. 54 is a perspective drawing showing the χ-plane and the ϕ-, ω-axes at the reflecting position.)

Let ρ be the fixed angle between the reciprocal lattice vector OP and the positive direction of the goniometer-head axis. ρ is related to the cylindrical co-ordinates ξ, ζ by the equation

$$\tan\rho = \xi/\zeta, \tag{2.26}$$

and we will assume for the present that it lies in the range

$$0 \leqslant \rho \leqslant 90°.$$

In the Napierian triangle BQT of Fig. 16b

$$QT = 90° - \rho, \quad Q\hat{B}T = -\chi, \quad BQ = 90° - \epsilon,$$

so that

$$\sin\chi = -\frac{\cos\rho}{\cos\epsilon}. \tag{2.27}$$

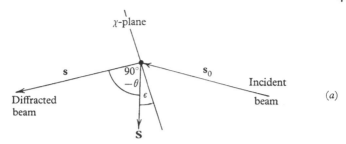

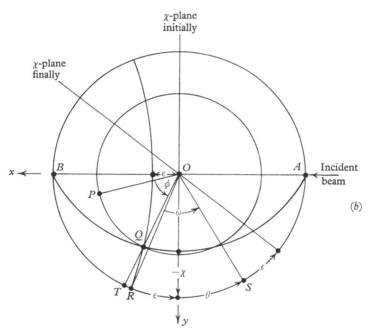

Fig. 16. Normal-beam equatorial method in general setting: (a) definition of off-set angle ϵ, which is positive when the χ-plane lies between the scattering vector **S** and the incident beam; (b) stereographic projection onto equatorial plane, showing relations between ϵ and the angles ω, χ, ϕ (after Willis, 1962 a).

Thus the reciprocal lattice vector, which is characterized by the angle ρ, can be brought into the reflecting position provided that ϵ lies in the range $\rho \geqslant \epsilon \geqslant -\rho$. Any value of ϵ can be chosen in this range, and the particular value selected determines the azimuthal orientation ψ of the reflecting plane.

The stereogram in Fig. 17 gives the relation between ϵ and ψ. *CED* represents the *(hkl)* plane after rotation by ϕ, and *Q* is the pole of the corresponding normal. The χ-rotation displaces *Q* to *R* and *C* to *G* along small circles normal to the *x*-axis *AB*. The inclination of the vector $\overrightarrow{OC}$ to the equatorial plane is not affected by the ϕ-rotation and the vector $\overrightarrow{OG}$ likewise is unaffected by the

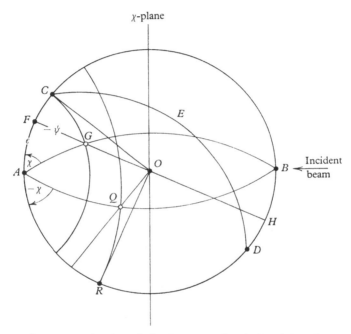

Fig. 17. Stereogram showing relation between azimuth ψ and the offset angle ϵ. *CED* is the reflecting plane before the χ-rotation and *FOH* its position after the χ-rotation (after Willis, 1962 *a*).

ω-rotation. Consequently, the azimuth ψ of the plane *CED* changes only during the rotation about the χ-axis. ψ is the angle of rotation of the crystal about the scattering vector **S**; we define it as positive for clockwise rotation looking along the positive direction of **S** and as zero when $\epsilon = 0$. In Fig. 17, ψ is equivalent to the change in inclination to the equatorial plane of the vector $\overrightarrow{OC}$ as it moves to the position $\overrightarrow{OG}$ during the rotation about the

χ-axis. Thus $FG = -\psi$, and from the Napierian triangle AFG, with $AF = \epsilon$ and $F\hat{A}G = \chi$, we have

$$-\tan\psi = \sin\epsilon\tan\chi. \tag{2.28}$$

Combining (2.27) and (2.28) gives

$$\tan\psi = \frac{\sin\epsilon\cos\rho}{(\cos^2\epsilon - \cos^2\rho)^{\frac{1}{2}}}. \tag{2.29}$$

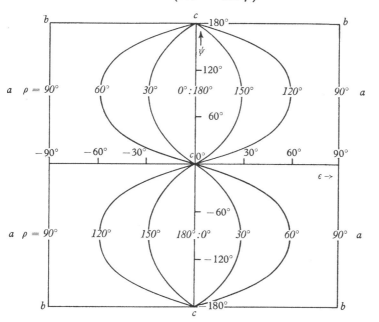

Fig. 18. Dependence of azimuth ψ on the offset angle ϵ for different values of ρ. ρ is the angle between the reciprocal lattice vector and the positive ϕ-axis. For a particular reflexion, characterized by a fixed value of ρ, the azimuth can be varied through 360° by varying ϵ in the range $-\rho \leqslant \epsilon \leqslant \rho$ for $\rho < 90°$, or in the range $-(180°-\rho) \leqslant \epsilon \leqslant 180°-\rho$ for $\rho > 90°$. The fixed-χ setting is represented by the horizontal lines aa and the symmetrical-B setting by the vertical lines bb. The symmetrical-A setting corresponds to the points c where all the curves converge.

Equation (2.29) is the required relation between the azimuth ψ and the offset angle ϵ of the χ-circle, for those reflecting planes inclined at an angle ρ to the ϕ-axis.

Fig. 18 shows graphically the relation between ψ and ϵ for the full range 0–360° in ψ. For reflexions with $\rho = 0$ (180°) or 90° the azimuth can be changed without altering the offset angle. If $\rho = 0\,(180°)$ the scattering vector lies along the goniometer-head axis and ψ can be made to vary continuously by changing ϕ, whereas if $\rho = 90°$ the scattering vector lies along the χ-axis and ψ can be varied continuously by changing χ. For intermediate values of ρ, ψ depends on the value chosen for the angle ϵ, and there are corresponding changes in the setting angles ω, χ, ϕ, which also depend on ϵ. We must now derive expressions for ω, χ, ϕ in terms of the offset angle and the co-ordinates ξ, ζ, τ.

General setting angles of crystal

ω, χ, ϕ are the setting angles required to bring the crystal from the standard orientation (defined on p. 38) to the reflecting position for the (hkl) plane.

ω is given from Fig. 16b as

$$\omega = \theta + \epsilon, \tag{2.30}$$

and χ is given by equation (2.27). To derive an expression for ϕ, we have from Fig. 16b:

$$\phi = B\hat{O}T - B\hat{O}P. \tag{2.31}$$

But $B\hat{O}P = \tau$, the angular polar co-ordinate of P, and $B\hat{O}T$ is related to the angles $-\chi$ and $90° - \epsilon$ in the Napierian triangle BQT by

$$\tan B\hat{O}T = \sin(90° + \chi)\cot\epsilon. \tag{2.32}$$

Combining (2.27), (2.31) and (2.32) gives:

$$\phi = \tan^{-1}\left(\frac{(\cos^2\epsilon - \cos^2\rho)^{\frac{1}{2}}}{\sin\epsilon}\right) - \tau. \tag{2.33}$$

The required equations for the setting angles are (2.27), (2.30) and (2.33).

Let us suppose that the offset angle is ϵ_0, where ϵ_0 is negative. Thus ϵ_0 lies in the range $0 \leqslant \epsilon_0 \leqslant -\rho$, and the hkl point is in the upper hemisphere of reciprocal space (that is, ρ lies between 0 and 90°). From equation (2.29) the azimuthal orientation ψ_0 of the reflecting plane lies between 0 and 90° and is given by

$$\tan\psi_0 = \frac{\sin\epsilon_0\cos\rho}{(\cos^2\epsilon_0 - \cos^2\rho)^{\frac{1}{2}}}.$$

The setting angles are

$$\omega = \theta + \epsilon_0,$$
$$\chi = -90° + \chi_0 \quad (0 \leqslant \chi_0 \leqslant \rho),$$

and

$$\phi = \phi_0 - \tau \quad (0 \leqslant \phi_0 \leqslant 90°), \qquad (2.34)$$

where

$$\cos\chi_0 = \frac{\cos\rho}{\cos\epsilon_0},$$

and

$$\tan\phi_0 = \frac{(\cos^2\epsilon_0 - \cos^2\rho)^{\frac{1}{2}}}{\sin\epsilon_0}.$$

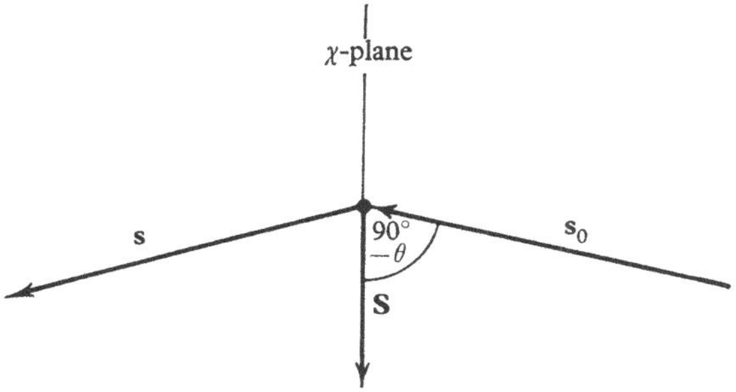

Fig. 19. Relation of χ-plane and scattering vector in symmetrical-A setting.

The expressions (2.34) for the setting angles can be extended to include the full range of azimuth from 0 to 360°. The results are presented in Table III.

Special settings in the equatorial method arise, when ϵ is assigned a particular value for each reflexion. Some of these special settings are discussed below.

Symmetrical-A setting ($\epsilon = 0$). At the reflecting position the χ-circle is symmetrically related to the incident and reflected beams (Fig. 19) and contains the scattering vector $\mathbf{S}$. The expressions for the setting angles in (2.30), (2.27) and (2.33) simplify to:

$$\omega = \theta,$$
$$\sin\chi = -\frac{\zeta}{(\xi^2 + \zeta^2)^{\frac{1}{2}}},$$

and

$$\phi = 90° - \tau. \qquad (2.35)$$

The detector is at an angle 2θ to the straight-through direction of the incident beam, so that the ω and 2θ shafts of the diffractometer can be geared together in a $1:2$ ratio, reducing the number of setting angles for the crystal and detector to three. Each reflexion is observed at only one azimuthal orientation of the reflecting plane, apart from the single plane normal to the ϕ-axis, for which any value of the azimuth ψ can be chosen by varying ϕ.

TABLE III. *Setting angles for different values of the azimuth ψ: equatorial geometry*

	$\psi = 0$	$\psi = \psi_0$	$\psi = 90°$	$\psi = 180° - \psi_0$	$\psi = 180°$
ϵ	0	ϵ_0	$-\rho$	ϵ_0	0
ω	θ	$\theta - |\epsilon_0|$	$\theta - \rho$	$\theta - |\epsilon_0|$	θ
χ	$-90° + \rho$	$-90° + \chi_0$	$-90°$	$-90° - \chi_0$	$-90° - \rho$
ϕ	$90° - \tau$	$\phi_0 - \tau$	$-\tau$	$-\phi_0 - \tau$	$-90° - \tau$
Type of setting:	Symmetrical-A	General	Fixed-χ	General	Symmetrical-A

	$\psi = 180° + \psi_0$	$\psi = 270°$	$\psi = 360° - \psi_0$	$\psi = 360°$
ϵ	$-\epsilon_0$	ρ	$-\epsilon_0$	0
ω	$\theta + |\epsilon_0|$	$\theta + \rho$	$\theta + |\epsilon_0|$	θ
χ	$-90° - \chi_0$	$-90°$	$-90° + \chi_0$	$-90° + \rho$
ϕ	$180° + \phi_0 - \tau$	$180° - \tau$	$180° - \phi_0 - \tau$	$90° - \tau$
Type of setting:	General	Fixed-χ	General	Symmetrical-A

In this table:

$$\sin \epsilon_0 = -\frac{\tan \psi_0 \sin \rho}{(\tan^2 \psi_0 + \cos^2 \rho)^{\frac{1}{2}}},$$

$$\cos \chi_0 = \cos \rho / \cos \epsilon_0,$$

and　　$\tan \phi_0 = (\cos^2 \epsilon_0 - \cos^2 \rho)^{\frac{1}{2}} / \sin \epsilon_0 \quad (0 \leqslant \phi_0 \leqslant 90°).$

ξ, ζ, τ are the cylindrical polar co-ordinates of the hkl point and $\tan \rho = \xi/\zeta$.
In the upper hemisphere of reciprocal space $(0 \leqslant \rho \leqslant 90°) 0 \leqslant \epsilon_0 \leqslant -\rho$ and $0 \leqslant \chi_0 \leqslant \rho$.
In the lower hemisphere of reciprocal space $(90° \leqslant \rho \leqslant 180°) 0 \leqslant \epsilon_0 \leqslant -180° + \rho$ and $0 \leqslant \chi_0 \leqslant 180° - \rho$.

In this setting all reflexions lying within the limiting sphere are, in theory, accessible. In practice, the χ-circle, if it is a complete circle, may obstruct the passage of the incident and reflected beams at high Bragg angles θ. For the symmetrical-B setting with $\epsilon = 90°$ (or $-90°$) there is no such difficulty in observing high-angle reflexions.

We show in §2.5 that the symmetrical-A setting corresponds to the equi-inclination setting of the inclination method, and so has similar limitations (for instance, with respect to the occurrence of simultaneous reflexions). The symmetrical-*A* setting is widely used for making intensity measurements, even though the general setting can be employed just as easily in a four-circle instrument. With the general setting the azimuthal orientation of the reflecting plane can be varied: we shall refer to the importance of this variation in Chapter 9.

Symmetrical-B setting ($\epsilon = 90°$). Here the χ-circle is symmetrically oriented with respect to the incident and diffracted beams, and is normal to the scattering vector **S** (Fig. 20). The *hkl* plane can be brought into the reflecting position provided it lies in a zone with the ϕ-axis as the zone axis. All other reflexions are

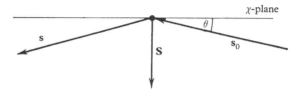

Fig. 20. Relation of χ-plane and scattering vector in symmetrical-*B* setting.

inaccessible. Thus, as for the anti-equi-inclination Weissenberg setting, the symmetrical-*B* setting is restricted to the collection of two-dimensional intensity data in the zero level.

Each accessible reflexion can be measured at any arbitrary value of the azimuth ψ. The reflecting normal lies along the χ-axis, and a 360° variation in ψ is achieved by rotating the crystal in the reflecting position about the χ-axis.

The setting angles ω, ϕ are

$$\left.\begin{aligned} \omega &= \theta + 90°, \\ \phi &= -\tau. \end{aligned}\right\} \tag{2.36}$$

Fixed-χ setting ($\epsilon = |\rho|$). If the offset angle ϵ is equal to the value of ρ for each reflexion to be measured, we have the so-called 'fixed-χ' setting. This setting has been discussed by Wooster & Wooster (1962).

Fig. 21 is the stereogram showing the setting angles required to bring the reciprocal lattice point P to the reflecting position at S. The setting angles are given by:

$$\left.\begin{aligned}
\omega &= \theta + \rho, \\
\chi &= -90°, \\
\phi &= -\tau.
\end{aligned}\right\} \tag{2.37}$$

Thus χ is fixed at $-90°$ and there are three variable setting angles in all, including 2θ for the detector. The fixed-χ setting is particularly useful where physical access to the crystal is required, as in

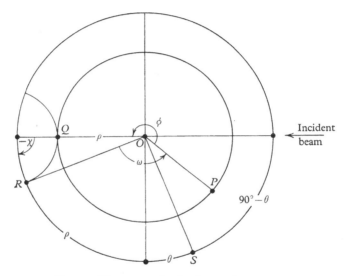

Fig. 21. Stereogram for fixed-χ setting.

the provision, for instance, of ancillary apparatus such as a furnace (see p. 81): the χ-circle is replaced by a simple bracket and so there is no obstruction to access to the crystal from above.

Table III summarizes the formulae given above for the setting angles in the general, symmetrical-A and fixed-χ settings: the table includes the full range of 360° in the azimuthal angle. By combining this table with Table I containing the formulae for the cylindrical polar co-ordinates ξ, ζ, τ, we obtain expressions for the setting angles ω, χ, ϕ in terms of the indices hkl and the lattice parameters.

Regions of physical interference

In the normal-beam equatorial geometry there are angular regions of the diffractometer in which no measurements can be made because of physical interference between parts of the instrument. These obstructions are particularly serious in X-ray diffractometers if long collimators are used which extend right up to the crystal position: such collimators are used to reduce air scattering (see p. 172). These obstructed regions are different in type from the 'blind regions' discussed in §2.3 for inclination geometry. The blind regions are a property of inclination geometry and they have no counterpart in equatorial geometry.

There are at least three kinds of obstructed regions. The first of these is a small cone-shaped region in which the incident or diffracted beam is shadowed by the goniometer head. The second kind arises from interference between the sides of the χ-circle and the detector collimator (Fig. 22). The result of this form of obstruction is that there is a range of inaccessible values of $|2\theta - \omega|$. The third kind of interference is between the source collimator and the χ-circle (Fig. 23): it can be reduced by making this circle an incomplete one (Eulerian cradle) at the expense of reducing the range of χ to less than a complete 360° rotation.

The plane of the χ-circle is frequently offset from the ω-axis: this procedure reduces obscuration of the incident beam and increases that of the diffracted beam, or vice versa, depending on whether the χ-circle is on the detector or the source side of the crystal (Fig. 22). It is frequently impossible to pass from one non-obscured region to another via an intervening obscured region without removing a collimator: this fact complicates the writing of setting programs.

In spite of the limitations described above there are, in practice, no blind regions in reciprocal space for a crystal examined with a four-circle diffractometer, even if the χ-circle is a complete circle. This is because of the extra rotational degree of freedom (see p. 40) given by this instrument. All reflexions can be observed up to a maximum scattering angle $2\theta_{max.}$ and the physical limitations can be considered as restrictions on the accessible range of the azimuth ψ for each reflexion. A practical procedure is to calculate

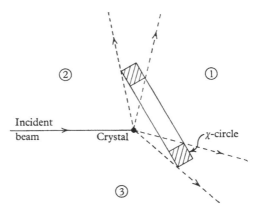

Fig. 22. Regions obstructed by the vertical circle, which is offset and viewed in projection on the equatorial plane. If the detector collimator is close to the crystal it is not possible to pass between regions ①, ② or ③ without removing the collimator. The broken lines represent limiting positions of the reflected beam. In this case the χ-circle is on the detector side of the crystal.

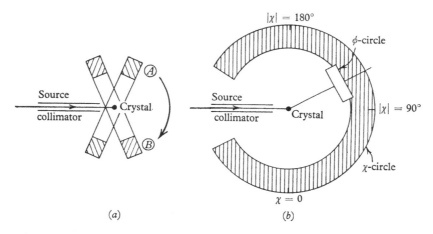

Fig. 23. An incomplete vertical circle viewed (a) in projection on the equatorial plane and (b) in elevation. It is possible to move from position Ⓐ to position Ⓑ without removing the source collimator. The $|\chi| = 90°$ position must be available, so that at least one arm of the vertical circle must extend above the equatorial plane.

the setting angles for a particular value of ψ using the formulae given in Table III: if this combination of setting angles is unacceptable because of physical interference, a new ψ is chosen and the calculation repeated until a satisfactory combination is found.

If χ is restricted to the range $0-90°$, as in the instruments described by Furnas & Harker (1955) and by Mayer (1964), we can only observe a hemisphere in reciprocal space on one side of the 'straight-through' position of the incident beam. To observe the other hemisphere we must either measure the reflexions on both sides of the incident beam (that is, θ both positive and negative), or rotate the χ-circle through 180° about the ω-axis: if neither of these alternatives is possible, the crystal must be re-orientated on the goniometer head between measuring each hemisphere.

Setting angles for general crystal orientation

The setting angles in Table III and the formulae for the cylindrical polar co-ordinates ξ, ζ, τ in Table I have been derived assuming that the **c**-axis of the crystal is along the goniometer-head axis ϕ and that the **a***-axis, at the standard orientation of the crystal, is along the incident beam direction. This implies that the crystal is mounted on goniometer arcs, which are adjusted to bring **c** into coincidence with the ϕ-axis.

However, goniometer arcs are not really necessary, in that the three circles ϕ, χ, ω allow the crystal to be rotated to any orientation in the laboratory space. The problem then remains of calculating the magnitudes of the setting angles when the crystal is mounted in any arbitrary orientation on the goniometer head. The appropriate formulae are given in a paper by Wooster (1965).

Powell (1966) has written a computer program in 'Fortran' which calculates the setting angles of a four-circle diffractometer for an arbitrary crystal orientation. The program has the following general features:

(i) A least-squares fitting procedure is used such that the precise orientation of the crystal with respect to the xyz laboratory co-ordinates is determined from the observed setting angles for a few hkl reflexions.

(ii) The azimuthal angle ψ is chosen so that unwanted reciprocal lattice points (giving simultaneous reflexions) are at least a certain distance from the Ewald sphere. This criterion defines a number of intervals in which ψ may lie, and the optimum angle is chosen to be at the mid-point of the longest interval.

(iii) Obstructed regions in reciprocal space, caused by the shape of the instrument obstructing the incident or diffracted beams, are avoided by expressing them as a number of inequality constraints; each constraint restricts one of the angles ω, ϕ, χ, 2θ and $\omega - 2\theta$, or involves logical 'and' combinations of pairs of these setting angles.

2.5. Comparison of inclination and equatorial methods

Table IV summarizes the properties of the inclination and equatorial methods, grouped under the special settings appropriate to each method. In X-ray work both methods are used, although the general inclination setting, requiring alteration of the inclination of both the X-ray tube and the detector to the goniometer-head axis, is not convenient for mechanical reasons. For neutron work the equatorial method is more suitable, as the detector requires heavy shielding and it is preferable to restrict its motion to a single rotational axis 2θ. Moreover, intensity errors arising from simultaneous reflexions are more troublesome in neutron diffraction, and by working in the general equatorial setting the azimuth ψ of the reflecting plane can be chosen to minimize these errors.

Because of differences in their mechanical construction we have described the geometries of inclination and four-circle diffractometers separately. However, the two geometries must be related, as they both achieve the same requirements, discussed in §2.1, for setting the crystal and detector and for measuring the reflexion. This correspondence was first pointed out by Phillips (1964), who discussed the correspondence between special settings used in the two geometries.

Fig. 24 shows the χ-plane and the equatorial plane in the general equatorial setting, and the two inclination angles μ, ν in the general inclination setting. The χ-plane is denoted by the great circle WUV, where WCV is a vertical diameter and CU is in the horizontal, equatorial plane. ACO is along the incident beam direction; CS is the bisector of the angle ACT, so that the angle

TABLE IV. *Summary of properties of inclination and equatorial methods*

Method	Setting angles		Special features and limitations
	Crystal	Detector	
Inclination:			
General	ϕ, μ	Υ, ν	Azimuth ψ varies with μ, but mechanically inconvenient to vary μ, ν together between reflexions
Normal-beam	ϕ $(\mu = 0)$	Υ, ν	Only two angles, ϕ and Υ, to be set within each reciprocal lattice level. Blind regions in upper levels. No choice of ψ
Equi-inclination	ϕ $(\mu = -\nu)$	Υ, ν	Only two angles to be set within each level. No blind regions. No choice of ψ
Anti-equi-inclination	ϕ $(\mu = \nu)$	Υ	One level only, $\zeta = 0$, accessible. This level can be measured at any $\mu \, (= \nu)$ and ψ is varied by changing μ
Flat-cone	ϕ, μ $(\nu = 0)$	Υ	Detector moves in one plane only. Blind regions for upper levels. No choice of ψ. Used for simultaneous measurement of several reflexions (see p. 56)
Equatorial:			
General	ω, χ, ϕ	2θ	Detector moves in equatorial plane. Each reflexion can be measured at any azimuth ψ
Symmetrical-A	ϕ, χ $(\omega = \theta)$	2θ	No choice of ψ
Symmetrical-B	ϕ, χ	2θ	One level only, $\zeta = 0$, is accessible. Each reflexion in this level can be measured at any ψ
Fixed-χ	ϕ, ω $(|\chi| = 90°)$	2θ	No choice of ψ. Crystal accessible for high- or low-temperature attachments
Flat-cone	ϕ, χ $(\omega = 2\theta)$	2θ	As for flat-cone setting in inclination method

$UCS = \epsilon$. The ϕ-axis, which lies in the χ-plane, is tilted forwards at an angle $WCB \, (= -\chi)$ to the vertical axis.

In the inclination method the angle between the incident beam and the positive goniometer-head axis is $90° + \mu$, and the angle between the positive goniometer-head axis and the reflected beam is $90° - \nu$. Thus

$$A\hat{C}B = 90° + \mu, \quad B\hat{C}T = 90° - \nu$$

and these angles are related to other angles in the Napierian triangles BUT, BUA, as shown in Fig. 25. Application of Napier's rules gives

$$\cot\omega = \frac{\cos 2\theta - (\sin\nu/\sin\mu)}{\sin 2\theta}. \qquad (2.38)$$

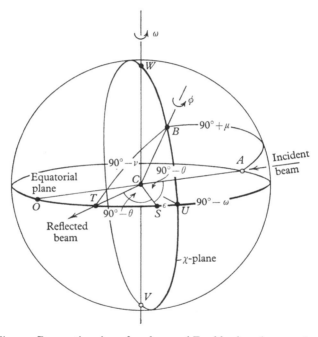

Fig. 24. Perspective view of χ-plane and Ewald sphere in general setting of normal-beam equatorial method.

We can apply this equation directly to the special settings of the inclination method:

(1) *Equi-inclination setting,* $\mu = -\nu$. From equation (2.38) $\omega = \theta$ and from (2.30) $\epsilon = 0$, so that this setting corresponds to the symmetrical-A setting of the equatorial method, in which the χ-circle bisects the angle between the incident and reflected beams.

(2) *Anti-equi-inclination setting,* $\mu = \nu$. From the same equations

$\omega = 90° + \theta$ and $\epsilon = 90°$, which corresponds to the symmetrical-B setting of the equatorial method.

(3) *Flat-cone setting*, $\nu = 0$. Here $\omega = 2\theta$ and $\epsilon = \theta$, that is, the axis of the χ-circle is parallel to the reflected beam.

(4) *Normal-beam setting*, $\mu = 0$. Here $\omega = 0$ and $\epsilon = -\theta$, that is, the axis of the χ-circle is parallel to the incident beam.

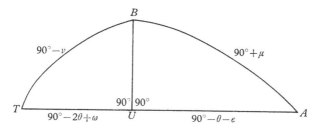

Fig. 25. Spherical triangles in Fig. 24 bounded by χ-plane and equatorial plane.

It is instructive to consider the correspondence between the special settings of the two methods in studying problems associated with the occurrence of simultaneous reflexions. Thus in the symmetrical-A setting of an equatorial instrument the conditions for simultaneous reflexions must be automatically satisfied if the goniometer-head axis is perpendicular to a set of planes in the reciprocal lattice, just as they are in an equi-inclination instrument.

An example where it is easier to prove a property of a special setting in one type of geometry, by considering its counterpart in the other type, is the following in the anti-equi-inclination setting a reflexion in the $\zeta = 0$ level can be observed at any inclination angle μ; because of the formal equivalence of this setting with the symmetrical-B setting, in which χ assumes any arbitrary value and provides a continuous variation of the azimuth ψ, this change in μ must also correspond to an alteration of ψ. However, because of mechanical limitations on μ, a complete variation of ψ is not possible in the anti-equi-inclination setting.

Of course, the correspondence between inclination and four-circle instruments is not complete. The ω- and χ-axes in the equatorial method have no mechanical counterpart in the inclination method, and the single rotation about the 2θ-axis in the equatorial method is resolved into component rotations about ν

and Υ in the inclination case. The main feature of the inclination method is that the inclination angles μ, ν remain constant for all reflexions within any one level of the reciprocal lattice, and the level can be surveyed by varying Υ and ϕ only. Measuring the reflexions level-by-level ceases to be the logical procedure in an equatorial instrument.

2.6. The simultaneous measurement of reflexions

The diffractometer methods which have been described so far have an obvious limitation when compared with photographic techniques: only one reflexion is recorded at a time even though the conditions may be satisfied for two or more reflexions to occur simultaneously. In crystals with very large unit cells, for instance protein or virus single crystals, several thousand reflexions may occur at once. This situation is illustrated by Plate I. Area detectors of the type discussed in Chapter 4 offer the possibility of recording these reflexions simultaneously; the development of such detectors is still in its infancy, but they may eventually give rise to a radically new concept of diffractometer design. Phillips (1964) has shown how a limited increase in the efficiency of existing single crystal diffractometers, based on either the inclination or equatorial geometry, is possible by measuring a few reflexions at a time.

The measurement of simultaneous reflexions in pairs

We have referred on p. 32 to the occurrence of double or triple reflexions, which arise from the simultaneous presence of two or three reciprocal lattice points on the surface of the Ewald sphere. The conditions for a reflexion to occur are satisfied when the corresponding reciprocal lattice point lies on the Ewald sphere. These conditions are still satisfied when the crystal is rotated about the normal to the (*hkl*) plane: during this rotation many other reciprocal lattice points pass through the sphere of reflexion, and repeatedly give rise to the conditions for the occurrence of simultaneous reflexions. The intensity of the *hkl* reflexion is affected by the simultaneous occurrence of a second reflexion, producing errors in the measurement of *hkl*. Fortunately, these errors are often small and there are many investigations where, far from having to

PLATE I

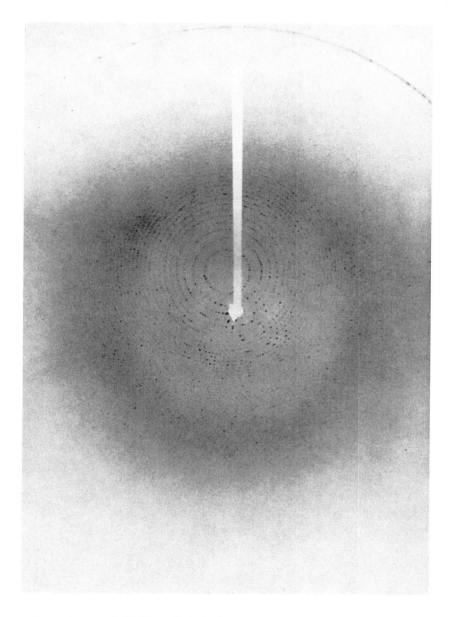

X-ray photograph (CuKα radiation) of *stationary* crystal of poliomyelitis virus
(Finch & Klug, 1959).

eliminate simultaneous reflexions, we can exploit their occurrence to speed up the collection of intensity data.

Fig. 26 shows the Ewald sphere in the general inclination setting, with a negative value of the inclination angle μ. Let us suppose that the crystal is mounted with a reciprocal-lattice axis, say $\mathbf{c}^*$, along the goniometer-head axis ϕ. (Note that this mounting of the crystal is different from that which we have discussed in earlier

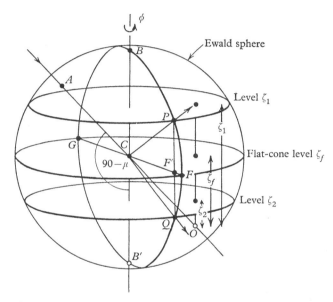

Fig. 26. Measurement of two reflexions P, Q simultaneously using the inclination geometry.

sections of this chapter, where we assumed that a crystal axis, $\mathbf{c}$, coincides with the goniometer-head axis.) Any pair of reciprocal lattice points with co-ordinates hkl_1 and hkl_2 will lie simultaneously on the surface of the Ewald sphere, if they are symmetrically related to the flat-cone level at $\zeta = \zeta_f$ passing through the centre of the sphere. Thus the points P and Q lying on levels ζ_1 and ζ_2 will give simultaneous reflexions if

$$\zeta_f = \tfrac{1}{2}(\zeta_1 + \zeta_2). \qquad (2.39)$$

The reflected beams CP, CQ lie in a plane $BPFQB'$ containing the rotation axis BB', to which they are inclined at angles $90° \pm \nu$, where

$$\sin\nu = \tfrac{1}{2}(\zeta_1 - \zeta_2). \tag{2.40}$$

The flat-cone level can either coincide with a possible reciprocal lattice level or lie half way between two such levels.

An inclination diffractometer can be adapted, therefore, for the simultaneous measurement of reflexions in pairs by providing it with two detector arms inclined at $\pm\nu$, where ν is given by (2.40). The instrument is set at the inclination angle μ for the flat-cone level—that is, $\sin\mu = -\zeta_f = -\tfrac{1}{2}(\zeta_1 + \zeta_2)$ from (2.39)—and any pair of reflexions hkl_1, hkl_2 with $\zeta = \zeta_1$ and $\zeta = \zeta_2$ can be measured simultaneously, provided they do not fall into the blind region which always exists for levels other than the equi-inclination (and anti-equi-inclination) levels.

Because of the correspondence between the general inclination and equatorial methods, discussed in §2.5, we expect that a similar procedure is possible for measuring two reflexions simultaneously using a four-circle diffractometer. Fig. 27 illustrates the situation for measuring the two reflexions CP and CQ in symmetrical positions above and below the equatorial plane $AUFOTG$, here viewed from below; the goniometer-head axis CB is drawn vertically to facilitate comparison with Fig. 26. The crystal is mounted with $\mathbf{c}^*$ along the goniometer-head axis, which moves round the χ-circle $BWTVU$. The great circle FGH represents the flat-cone level at ζ_f with the point P ($= hkl_1$) at $\Delta\zeta$ above the flat-cone level and the point Q ($= hkl_2$) at $\Delta\zeta$ below; both P and Q lie simultaneously on the Ewald sphere and the line PQ is parallel to $\mathbf{c}^*$. To measure P and Q together two detectors are mounted on the detector arm CF. These detectors are inclined at $\pm\nu$ to CF, where

$$\sin\nu = \Delta\zeta = \tfrac{1}{2}(\zeta_1 - \zeta_2).$$

The detectors lie in a plane which contains the $\mathbf{c}^*$-axis and is at right-angles to the χ-plane, in accordance with the condition $\omega = 2\theta$ for the flat-cone setting (§2.5). All reflexions of the type P, Q are measured in pairs which have a fixed difference of $2\Delta\zeta$ in their ζ co-ordinates. It is only necessary to set three independent angles ω, χ, ϕ for each pair, as the detector arm at the setting angle 2θ is permanently locked to the ω-shaft in order to keep the arm normal to the χ-plane.

Quasi-simultaneous reflexions

It is customary to measure integrated intensities by recording the number of diffracted quanta as the crystal is rocked through the reflecting position. The rocking range must be sufficient to allow all parts of the crystal to reflect radiation from all parts of the X-ray tube focus, and the range depends on a number of factors, such as mosaic spread of the crystal, divergence of the

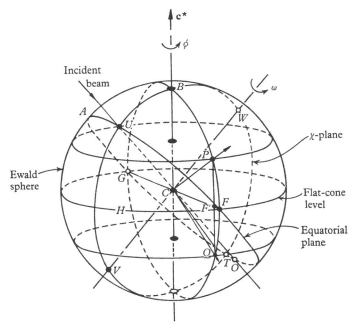

Fig. 27. Measurement of two reflexions P, Q simultaneously using the equatorial geometry.

incident beam and the range of incident wavelengths. For many investigations a rocking range of 1–2° is adequate. Phillips (1964) has shown that when the spacing between reciprocal-lattice levels is small enough, reflexions from a number of levels can appear quasi-simultaneously, that is, during the 1–2° rocking range of the crystal, even though the corresponding reciprocal-lattice points do not touch the Ewald sphere together.

Referring to Figs. 26 and 27 we shall assume that the points P, F', Q represent three reciprocal-lattice points in adjacent levels. Thus their ξ co-ordinates are all equal and their ζ co-ordinates are $\zeta_f + \Delta\zeta$, ζ_f, $\zeta_f - \Delta\zeta$, respectively. P and Q touch the Ewald sphere together, but at a different time from F', and we shall calculate the

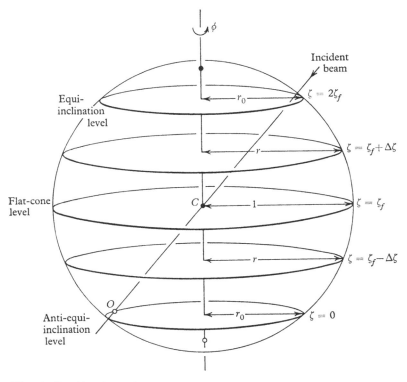

Fig. 28. Ewald sphere showing general levels at $\zeta = \zeta_f + \Delta\zeta$ symmetrically related to the flat-cone level at $\zeta = \zeta_f$. O is the origin of reciprocal space and C the centre of the Ewald sphere.

difference $\Delta\phi$ in the setting angle of the crystal between the measurement of F' and of P and Q.

Fig. 28 is a view of the Ewald sphere with the crystal set for the non-zero level ζ_f to be recorded in the flat-cone setting. The levels $\zeta_f + \Delta\zeta$, $\zeta_f - \Delta\zeta$ are also shown. The corresponding circles of reflexion are illustrated in Fig. 29, together with the angle $\Delta\phi$ and the points F' and P at the reflecting position. The radii of the various circles of reflexion are given by

$$r_0^2 = 1 - \zeta_f^2 \quad \text{for the equi-inclination level } (\zeta = 2\zeta_f),$$

and $\quad r^2 = 1 - \Delta\zeta^2 \quad$ for the $\zeta_f \pm \Delta\zeta$ levels,

while the flat-cone level has unit radius.

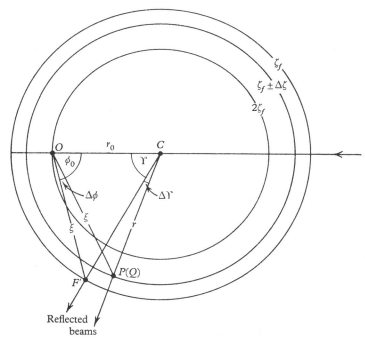

Fig. 29. Projection of Fig. 28 down ζ-axis, showing reflected beams for reciprocal lattice points F', $P(Q)$ in levels $\zeta = \zeta_f$, $\zeta_f \pm \Delta\zeta$ and at distance ξ from crystal rotation axis (after Phillips, 1964).

We have from Fig. 29:

$$\cos\phi_0 = \frac{\xi^2 + r_0^2 - 1}{2\xi r_0},$$

and

$$\cos(\phi_0 - |\Delta\phi|) = \frac{\xi^2 + r_0^2 - r^2}{2\xi r_0},$$

so that

$$\cos\phi_0 - \cos(\phi_0 - |\Delta\phi|) = \frac{r^2 - 1}{2\xi r_0}. \tag{2.41}$$

For small values of $\Delta\phi$ equations (2.41) can be rewritten

$$\Delta\phi = \frac{r^2 - 1}{2\xi r_0 \sin\phi_0},$$

or

$$\Delta\phi = -\frac{\Delta\zeta^2}{2\xi(1 - \zeta_f^2)^{\frac{1}{2}}\sin\phi_0},$$

where

$$\cos\phi_0 = \frac{\xi^2 - \zeta_f^2}{2\xi(1 - \zeta_f^2)^{\frac{1}{2}}}.$$

(2.42)

Fig. 30 has been calculated from equation (2.42) and shows how $\Delta\phi$ varies with ξ and $\Delta\zeta$ for either an inclination or a four-circle diffractometer in the flat-cone setting. Each line on this figure refers to a different value of ζ_f for the flat-cone level and is a contour within which $\Delta\phi \leqslant 0\cdot2°$. As noted in an earlier section, there is a blind region whose area increases with ζ_f, the minimum observable value of ξ being $1 - (1 - \zeta_f^2)^{\frac{1}{2}}$.

The separation of levels, $2\Delta\zeta$, in which reflexions can be measured at the same time as those in the flat-cone setting may be rather larger than Fig. 30 suggests. Equation (2.42) shows that $\Delta\phi$ is always negative with respect to the flat-cone level. If ϕ is set at $+0\cdot2°$ from its correct value for the flat-cone level, a value of $\Delta\phi = -0\cdot4°$ can be tolerated, since the reflexions in the flat-cone and the neighbouring levels would each be displaced by only $0\cdot2°$ from the centre of the rocking range. Assuming that $\Delta\phi \leqslant 0\cdot4°$ is the maximum allowable deviation in the setting angle, the limiting observable values of ξ can be calculated. These are plotted as full lines in Fig. 31 for the complete range of ζ_f from $\zeta_f = 0$ to $\zeta_f = 1\cdot0$ r.l.u., and for $\Delta\zeta$ up to $0\cdot08$ r.l.u. For moderate values of $\Delta\zeta$ the condition $\Delta\phi \leqslant 0\cdot4°$ imposes only a small addition to the blind region at the centre of the flat-cone level. Thus the additional blind region within which $\Delta\phi > 0\cdot4°$ comprises only $0\cdot7$ per cent of the accessible reflexions near $\zeta_f = 0$ when $\Delta\zeta = 0\cdot04$ r.l.u.; the corresponding figure for $\zeta_f = 0\cdot5$ r.l.u. is $0\cdot9$ per cent.

An additional restriction on the accessible range of reflexions is also caused by the change $\Delta\Upsilon$, shown in Fig. 29, of the setting angle of the detector. The maximum allowable value of $\Delta\Upsilon$ is determined by the maximum possible width of the detector aperture. The restriction imposed by $\Delta\Upsilon$ is only important at high values of ξ;

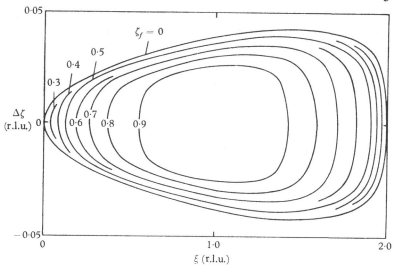

Fig. 30. Regions of reciprocal space within which $\Delta\phi \leqslant 0.2°$. The numbers 0, 0.3, 0.4, ..., 0.9 on the lines indicate the values of ζ_f for the flat-cone setting (after Phillips, 1964).

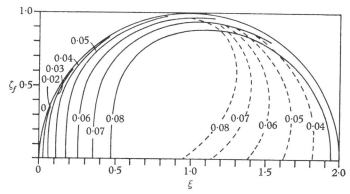

Fig. 31. Maximum and minimum values of ξ near the flat-cone setting when $\Delta\zeta = 0$ to 0.08 r.l.u. Full lines show limitations due to $\Delta\phi \leqslant 0.4°$, broken lines those due to $|\Delta\Upsilon| < 0.1°$ (after Phillips, 1964).

this is illustrated by Fig. 31 in which the broken lines, corresponding to different values of $\Delta\zeta$, are the boundaries of regions within which $|\Delta\Upsilon|$ is less than 0.1°.

Arndt, North & Phillips (1964) have described the adaptation of a linear diffractometer to permit the quasi-simultaneous measurement of three reciprocal lattice levels near the flat-cone setting. The single detector of the original instrument was replaced by a group of three detectors with their windows vertically above each other and in a line parallel to the goniometer-head axis. The detectors were equally spaced and ν was changed by mounting the detectors on a draw-tube, which allowed variation in the angles subtended at the crystal by the detectors. Separate counting circuits were provided for each detector, and the results recorded sequentially by means of a printing-out programmed scaler. Data collection rates of 3,000 reflexions/day have been achieved with this arrangement.

CHAPTER 3

THE DESIGN OF DIFFRACTOMETERS

3.1. Introduction

There is no one ideal X-ray or neutron diffractometer which is equally well suited to the many different investigations which have to be carried out. The principal factors in the performance of a diffractometer are:

> Accuracy of intensity measurement
> Speed of operation
> Number of accessible reflexions
> Amount of manual intervention required
> Accessibility of specimen
> Availability of a computer
> Versatility
> Reliability
> Cost

These varying requirements frequently conflict with one another and the choice of instrument for a given application will depend on the relative importance attached to them. We shall compare later the performance of two specific instruments under the headings listed above.

As we have seen in Chapter 2, diffractometers can be constructed according to two different geometrical arrangements, leading to a division into equatorial and inclination instruments. All diffractometers suitable for the collection of three-dimensional intensity data must have a number of shafts capable of being set independently: in the two arrangements the rotational degrees of freedom are allocated differently between crystal and detector shafts.

The general design of the diffractometer is dictated primarily by the type of geometrical arrangement, and by the method adopted in setting the shafts. The collimator, the goniometer head which supports the crystal, alinement aids such as viewing telescopes and even the detector itself can be thought of as exchangeable attachments which can be selected in accordance with the particular investigation in hand, and are of secondary importance in influencing the design of the instrument.

Classification of diffractometers according to shaft-setting methods

In any diffractometer, before a reflexion can be measured, the crystal must be oriented so as to bring the desired reciprocal lattice point upon the sphere of reflexion, and the detector, which is constrained to point towards the crystal, must be moved so as to pick up the reflexion. This setting operation is independent of the measurement of the reflexion, which involves the determination of the integrated intensity and of the background near the Bragg peak.

Four general setting procedures are possible, and these lead to a general classification of diffractometers into four groups. Only the last two groups will be discussed in detail.

(1) The shafts of the instrument may be set by hand to pre-computed values. The settings will generally be tabulated by an electronic computer; alternatively, special nomograms or analogue computers can be constructed for this purpose (Evans, 1953; Arndt & Phillips, 1957; Brown, 1958). It should be noted that when a diffractometer is hand-operated a much lower accuracy can be tolerated for the setting angles. It is a simple matter to 'tune-in' on a reflexion by giving small trial rotations to the shafts of the diffractometer while observing the diffracted intensity on a counting-rate meter. (Even in an automatic diffractometer it may be necessary to hunt for the peak of the reflexion, see p. 232. The ease of tuning-in is almost the sole advantage of manually operated instruments. While much valuable work has been done with non-automatic X-ray and neutron diffractometers, manually operated instruments represent only an interim stage; in the long run, a large degree of automation is essential if more than a very small volume of work is to be carried out.

(2) It is possible to change the crystal and detector settings continuously and automatically in such a way that all possible combinations occur in turn, a special measuring procedure being invoked whenever a Bragg reflexion is detected. In principle, no knowledge of the unit cell parameters or even of the initial

orientation of the crystal need be assumed if a search is made of the whole of reciprocal space for the Bragg reflexions. The only practical instrument of this type which has been constructed is an X-ray diffractometer of an equi-inclination type (Bond, 1955; Benedict, 1955). It was set manually for each reciprocal lattice level in turn, and the automatic search for reflexions was made within the level, which was surveyed in a spiral scan. Even then it was found that the complete coverage of the levels was very un-economical in time; although at least one structure determination has been reported (Geller & Katz, 1962) in which the data were collected on this instrument, it has not been developed further.

(3) The shafts of an inclination or an equatorial diffractometer may be set by methods akin to any of those employed for the automatic control of machine tools. The 'settings' are computed on a digital computer and presented on an output medium such as punched-paper tape or cards which can be read directly by the control circuitry of the diffractometer. These digital shaft-setting methods are discussed in the next section. As an alternative to the employment of an off-line computer and an auxiliary medium such as punched tape, the diffractometer may be connected directly, on-line, to the computer. The settings, and any other instructions necessary for complete automatic operation of the diffractometer, are read directly from the store of the computer into the diffracto-meter circuits. On-line operation is considered in Chapter 12.

(4) Instead of using a digital computer for the calculation of the crystal and detector shaft settings, these may be computed by analogue methods using specially designed mechanical or electrical devices which are integral parts of the diffractometer. In principle, there is not a large difference between the calculation of settings by an analogue method or by means of a digital computer; in practice, analogue devices impose certain restrictions on the design of the instrument as a whole, and it is convenient to consider such diffractometers separately. They are discussed below in §3.3.

3.2. Digital shaft setting

Any shaft-setting device consists of three parts:

There must be a store capable of receiving from the input device, and of holding in digital form, either the absolute position which the shaft must reach, or the increment, positive or negative, over the previous setting. Depending upon the nature of the information held in the store, the shaft-setting device as a whole is described as being absolute or incremental. Associated with the store there must be circuits which compare the actual and the desired positions of the shaft, and which initiate the appropriate rotations.

Attached to the shaft there must be a digitizer or encoder which converts the shaft movement or the shaft position into digital form. The digitizer indicates the absolute instantaneous position of the shaft or it produces signals which indicate how far the shaft has moved.

Finally, there must be a driving motor, together with its associated circuitry.

The store

During the reading-in of the setting information, successive digits of the number which represents the desired shaft angle are presented to the store in sequence; in the reprehensible jargon of the computer engineer they are 'staticized' in the store which then holds the whole of the number. When the shaft starts to move, a running comparison is carried out between the stored number and that currently derived from the digitizers, and the motor is brought to rest when coincidence is reached. In incremental shaft-setting operations the comparison and coincidence detection are replaced by a count-down to zero of the stored number. The store then becomes a pre-settable counter.

Various forms of stores can be used: they may be electro-mechanical devices such as relays or telephone-type stepping switches, bistable circuits (flip-flops), or core-stores. It is probably wiser to avoid electro-mechanical devices both because of their lower reliability and greater difficulty of servicing and because they limit the speed at which data are read into the store.

When the information is held in certain types of stores, such as stepping switches or magnetic cores, it is not destroyed by a momentary power failure or by switching off the installation. The numbers stored in bi-stables, on the other hand, are ephemeral: after an interruption of the supply the knowledge of the shaft position may be lost.

The general strategy of setting the shafts of a diffractometer is determined by whether an absolute or an incremental method of shaft-setting is adopted. In the former method each separate setting operation is independent of any preceding ones; in the latter any setting errors are cumulative and a fault which arises during one setting operation will invalidate all further operations. This property of incremental systems is not necessarily a disadvantage in diffractometers: it may be quite impossible to notice if one reflexion has been missed or wrongly measured because of a faulty setting, but a string of such errors would quickly manifest itself.

The incidence of errors with incremental shaft-setting devices is much reduced if some method is provided for returning the shafts to a datum point which can be reached without reference to the store or the digitizer. Such a datum point may be defined by a simple arrangement of micro-switches or by some more elaborate method: it is a great convenience if the datum points of the various shafts are mechanically adjustable. If such provision is made, a once-computed setting tape or deck of cards can be used for any initial orientation of the crystal about the goniometer-head axis, or for any zero position of the detector shaft. This zero position is defined by the 'straight-through' direction of the incident beam; it may change slightly after realining the diffractometer.

When an absolute shaft-positioning method is used, a separate store or register must be provided for each shaft, making it possible for all shafts to be positioned simultaneously. On the other hand, an incremental register can be shared between all the shafts of the instrument, since it is not required to retain any information after any one shaft has been set: the various shafts must then be positioned one at a time. This latter system may be the preferred one when the diffractometer contains three or more shafts: it may even be possible to share the basic circuitry between a number of diffractometers (Arndt & Willis, 1963 a). The absolute system is particularly suitable when only two shafts need be set automatically, as in an inclination diffractometer in which the data are collected level by level.

Digitizers

Many different kinds of digitizers have been devised, but basically they all belong to one of two classes, coded discs or pulse generators.

The principle of the coded disc is illustrated by Fig. 32, which represents a digitizer whose output is in the form of a six-bit binary number. The black and white parts of the diagram are opaque and transparent fields arranged on six tracks. Six photo-cells *A–F* receive light transmitted through the disc. If transmitted light represents a binary 1 and blocked light a binary zero, the reading at the point illustrated in the diagram is 001011. Actual

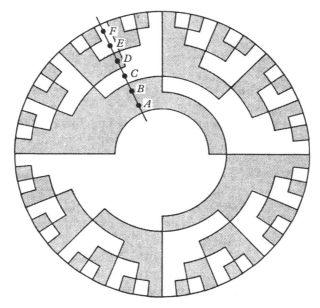

Fig. 32. Coded disc digitizer. *A–F* represent six photo-cells and the reading is 001011.

discs generally use a 'reflected code' which is somewhat different from the pure binary code illustrated; the coding arrangement is designed to avoid errors greater than one bit, which occur when the disc comes to rest with the division between an opaque and a transparent field immediately opposite a sensing element (see, for example, Evans, 1961). Electro-mechanical, capacitive or inductive coded discs may be used in place of optical discs.

Encoders or digitizers of this type have the advantage that they give a non-ephemeral absolute indication of the shaft position; their disadvantages are all traceable to the limited resolution obtainable with any but an exceedingly bulky encoder. Thus commonly available single-disc devices do not generally divide a complete revolution into more than 4,096 parts, or a little less than 0·1°. This resolution is too coarse to allow the discs to be mounted directly on the final diffractometer shafts, which are generally required to be set to a precision of about 0·01° or 0·02°. A practical encoder must either contain several discs geared together or be driven by intermediate shafts. In either case considerable demands are made on the accuracy of the gearing, and the encoder will be subject to wear when shaft setting is required at relatively high speeds.

In the second class of digitizer, the pulse generator, an electrical pulse is produced on one of two output lines. The single pulse corresponds to one unit of rotation of the shaft: if the rotation is in a positive direction the pulses appear on one output line, and if the direction is negative they appear on the other. These output lines are connected to the 'add' or 'subtract' inputs of bi-directional electronic counters. An example of a digitizing system of this kind is the moiré fringe measuring system (Guild, 1956), as developed by Ferranti Ltd. (Williamson, 1960). This system is adopted in the four-circle diffractometer described in §3.8. Pulse-generating setting methods generally allow higher setting speeds than other methods; the digitizer and pick-up device can often be made very compact. The principal disadvantage of these methods is that the absolute position of the shafts is recorded only in the reversible counters and not in a device attached to the shaft; the information is lost when power is disconnected.

The motor drive

The motor drive systems should be capable of moving the crystal and counter shafts of the diffractometer at high speeds. Slewing speeds of some ten to twenty degrees per second are desirable for X-ray diffractometers in order to allow complete freedom in the order in which reflexions are surveyed. With slower speeds some form of optimum programming is necessary in order to keep the angles moved through from one reflexion to the next as small as possible. Neutron diffractometers, which have detectors surrounded by massive shielding, may not have slewing speeds as high as those possible with X-ray diffractometers; the need for high speed is less in neutron diffraction, since the measuring time is longer and a little more time spent on the setting operation does not appreciably increase the total time for data collection.

When approaching the final setting, the shafts must decelerate gradually: this is to avoid any movement of the delicately mounted crystal.

The shafts must come to rest at the desired position without undue hunting; a single overshoot with a creep back to the rest position can usually be achieved. Clamping of the shafts at the correct position may be necessary to avoid subsequent movements under gravity.

The shafts can be set not only by d.c. servo-motors but also by multi-pole-stator impulse motors (Heaton & Mueller, 1960; Ladell & Cath, 1963; Arndt, Jones & Long, 1965), such as the American 'Slosyn' stepping motor or the British M-type motor. The rotors of these motors are indexed through a small fraction of a revolution by every pulse applied to its driving circuit. As d.c. power continues to be applied to the coil-driving circuit when the train of pulses ceases, these motors are subjected to a very strong braking torque which makes them virtually dead-beat. It should be noted, however, that not all these motors are capable of driving diffractometer shafts at the high slewing speeds demanded above.

Certain impulse motors index so accurately that a digitizer is not required: it is merely necessary to count the number of pulses fed into the driving circuit. A system of this kind behaves, of course, as an incremental setting system and the usual precautions of periodically checking for faults by a programmed datum return are necessary. In addition, as with certain types of digitizers, the accuracy of the device depends on the precision and freedom from backlash of the gear train.

One of the main drive motors can often be used to produce the small oscillation of the crystal required during the measuring procedure. Alternatively, an auxiliary oscillation mechanism driven by an additional motor may be needed (Arndt, Faulkner & Phillips, 1960; Arndt & Willis, 1963b).

3.3. Analogue shaft setting

In Chapter 2 we discussed the trigonometrical relations between reciprocal lattice co-ordinates and shaft-setting angles. It has been assumed so far that these equations will be solved by digital methods. The required inverse trigonometrical functions can also be evaluated with the aid either of sine-cosine resolvers (potentiometers) or of mechanical linkages. When such an analogue device is an integral part of the diffractometer, the instrument is generally, even if not with complete correctness, referred to as an analogue diffractometer.

Only one X-ray diffractometer has been described which makes use of an electrical analogue computer. This is the instrument designed by Drenck, Pepinsky & Diamant (1957). It is a conventional equi-inclination instrument in which the shaft angles μ, ϕ and Υ are evaluated by electrical analogue computers directly coupled to the shafts. No full description of this very interesting instrument appears to have been published.

All other analogue diffractometers which have been described employ mechanical analogues of the Ewald construction for the solution of Bragg's law.

Mechanical analogue diffractometers

The principle of the mechanical analogue diffractometer is illustrated in Fig. 33. Here XPO represents a horizontal section through the Ewald sphere with centre C, and XO is the incident beam, where O is the origin of the reciprocal lattice. In Fig. 33a, BOP is a central reciprocal lattice row and CP represents the direction of the diffracted beam when a reciprocal lattice point P lies on the circumference of the reflecting circle. Now suppose that BOP becomes an actual mechanical member pivoted at O, with a carriage P capable of motion along BOP: another bar of fixed length is pivoted at C and on the carriage P, and on this bar is fixed a radiation detector pointing at C. CD is a member supporting the crystal at C in such an orientation that a reciprocal lattice axis is parallel to CD, and the member CD is constrained to remain parallel to BOP by means of a parallelogram linkage or some other similar device. As the carriage P is moved along the slide BOP

successive reciprocal lattice points along the slide will be brought into the reflecting position on the circumference of the circle, and the resulting diffracted beams will be detected by the counter on *CP*.

So far the device is no more than an angle-bisecting mechanism which makes the crystal-carrying bar *CD* bisect the angle between

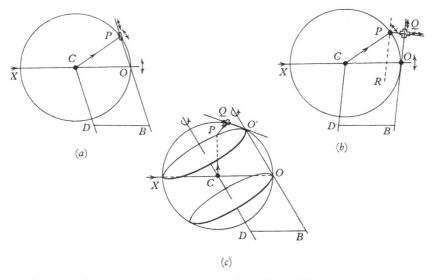

Fig. 33. Linkage system in a mechanical analogue linear diffractometer. (*a*) Scan of the central reciprocal-lattice row *h*oo (*POB*): a reflexion occurs whenever $OP = ha^*$. The carriage *P* moves along *POB*; the counter arm *CP*, assumed to be of unit length, is pivoted at *C* and on the carriage *P*; *POB* is pivoted at *O*; the crystal arm *CD* is kept parallel to *POB*. (*b*) Scan of a non-central reciprocal-lattice row *hk*o (*PR*); $PQ = kb^* = $ constant for the row as *Q* moves along *QOB*. (*c*) Measurement of an upper-level reflexion *hkl* (*P*) on the equi-inclination level *XPO'* with $OO' = lc^*$. As $O'Q = ha^*$ and $QP = kb^*$ are changed, *P* moves around the circle *XPO'*; *CD* is kept parallel to *OO'* and inclined at an angle $\sin^{-1}lc^*/2$ to the horizontal; the rotation about the axis *CD* is kept equal to the rotation about *OO'* (Arndt, 1964).

the incident and the diffracted beams. This simple arrangement can be used only for scanning along central reciprocal lattice rows; it can, however, readily be extended to the exploration of non-central reciprocal lattice rows such as *PR* in Fig. 33*b* by providing a cross-slide at *Q*. In this diagram, if the crystal is imagined so oriented that **a*** is parallel to *BOQ* and **b*** parallel to *PQ* (the angle *PQO* will of course not be a right angle for non-orthogonal levels), and if the length *PQ* is made equal to kb^* times the radius *CP*, then a movement of *Q* along *BOQ* will cause all the *hk*o reflexions to be scanned in succession. By adjusting the length of *PQ* after each traverse along *BOQ*, the zero reciprocal lattice level can be scanned in a zig-zag fashion.

The principles illustrated in Figs. 33*a* and *b* for the zero level can be extended to upper levels in one of two ways. Fig. 33*c* shows how a third slide *OO'* may be provided together with a second linkage which ensures equal rotations about *CD* and *O'B*. This is the method adopted by Arndt & Phillips (1961); it facilitates the setting of the instrument for upper levels but tends to impose restrictions on the maximum Bragg angle which can be achieved. Alternatively, as suggested by Mathieson (1958), the instrument can be tilted, as in Fig. 33*c*, so as to lie in the equi-inclination setting: the section of the Ewald sphere *XPO'* can then be treated as though it were the zero level and explored with the help of only two slides. The appropriate reciprocal lattice co-ordinates must be modified by a scale factor $\cos\mu$, which is the ratio of the radius of the circle *XPO'* to that of the Ewald sphere, where μ is the equi-inclination angle.

Ladell & Lowitzsch (1960) have pointed out that the movement of the carriages can be produced in one of three ways: by driving the crystal arm and thus forcing the analogue reciprocal axis system constituted by the cross-slide system and the counter arm to follow, by driving the counter arm, or by driving the carriages in a linear fashion along the slides. The first of these methods was used by Mathieson (1958), the second by Ladell & Lowitzsch (1960) in their diffractometer 'Pailred' I, and the third by Arndt & Phillips (1959, 1961) and by Ladell, Parrish & Spielberg in 'Pailred' II (1963). The third method, that of linear tracking in reciprocal space, has proved the most convenient and has given rise to the name 'linear diffractometer' for an instrument of this type.

In all linear diffractometers it is necessary to separate the setting operation from the crystal movement required during the measurement of the reflexion. A constant-speed linear motion along the slide produces a very non-uniform speed of rotation of the crystal about its axis. Consequently, if the slide motors were used for the measuring scan through the reciprocal lattice point, the Lorentz factor would have a highly inconvenient form. (The Lorentz factor expresses the relative time spent by each lattice plane in the reflecting position, and it is required in reducing the integrated intensities to relative structure factors (p. 278).) It is thus more convenient to effect the crystal oscillation during the measuring cycle by a separate mechanism which oscillates the crystal about the goniometer-head axis (CD in Fig. 33c).

In addition to the analogue diffractometers mentioned above, Potter (1962) has described an X-ray diffractometer operating on an analogue principle; improvements to this instrument have been discussed by Binns (1964).

3.4. Radiation shielding

The first requirement of the radiation shielding is to protect the operator from radiation hazards. Every diffractometer operator should be familiar with these hazards and with the precautions which should be taken (*International Tables for X-ray Crystallography*, vol. III, 1962). X-ray crystallographers are normally responsible for the whole of their equipment and for its safety aspects: in spite of this, they frequently seem less aware of the dangers than workers in the neutron field, who employ biological shielding surrounding their reactor and collimator assembly that has been carefully designed and tested. The commonest source of stray X-radiation is at the junction between the X-ray tube and the primary beam collimator. The provision of a ray-proof labyrinth at this point and of a fail-safe shutter is essential. Several satisfactory designs have been described (see Barnes & Franks, 1962). An adequate primary beam stop will further protect the operator from the primary beam. An X-ray diffractometer normally has a relatively open structure as compared with an X-ray camera in which all radiation scattered by the specimen and the beam may be intercepted by the film cassette; additional shielding should be installed to surround the diffractometer whenever the X-ray shutter is open. Especially when MoKα or AgKα radiations are employed, the only really safe shield is a cover with lead-glass windows which totally encloses the diffractometer. The most serious dangers arise during alinement of the collimators: interlocks should be provided which make it impossible for the shutter to be opened while the X-ray tube is switched on, unless the collimator is correctly seated in the radiation labyrinth. The adjustment of the specimen on its goniometer head is another operation in which there is great danger to the operator's fingers. However awkward and time-consuming it may be, this operation should only be carried out with the help of long wrenches or keys which will keep the fingers out of the beam.

A second requirement of radiation shielding concerns the detector and arises principally in neutron diffractometers. These may have to operate in the presence of fast neutron and γ-ray backgrounds. The detector must be surrounded, therefore, by a heavy shield (p. 139) to reduce the background counting rate. The complete detector assembly may weigh up to 30 kg and the detector arm must be sufficiently sturdy to support this weight. Neutron diffractometers usually employ normal-beam equatorial geometry in order to keep the motion of the heavy detector arm as simple as possible.

3.5. Collimator alinement

X-ray diffractometers have two collimators, a primary beam or source collimator through which the incident radiation travels from the source to the crystal, and a detector collimator through which the diffracted radiation passes to the detector (Fig. 34). The dimensions of the primary beam collimator which ensure optimum beam cross-section and divergence will be discussed in Chapter 6: in the present section we are concerned only with the need for the collimators to point accurately at the crystal and to lie exactly in the planes required by the diffractometer geometry.

The source collimator is usually mounted in a holder which is attached to the main base of the instrument. This holder must allow fine adjustment of both tilt and translation in the horizontal and vertical planes (Fig. 35). The adjusting screws are conveniently set by substituting for the collimator a steel rod of the same diameter turned to a sharp point. The point of the rod is brought into coincidence with the point of a needle centred on the goniometer head: with the aid of the crystal-viewing microscope it is possible to judge this coincidence to within 0·05 mm. The rod can be levelled to better than 0·01° using a high-accuracy spirit level: the diffractometer base should have levelling feet and machined reference surfaces, which can initially be made exactly horizontal. The detector collimator, which is mounted on the detector arm, is alined in a similar way.

Adjustments must be provided to move the diffractometer with respect to the X-ray tube, or vice versa, so that the source collimator points towards the centre of the focal spot. A foreshortened

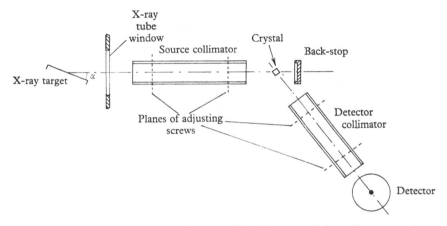

Fig. 34. Source and detector collimators. The diameters of the collimators and the magnitude of the take-off angle α are much exaggerated.

aspect of the line focus is viewed (Fig. 34), and it is convenient to be able to vary the take-off angle α between zero and about 6°. A very fine foreshortened focus can then be used for crystal aline-ment and unit-cell parameter measurement, a medium-sized focus for normal intensity measurements, and a broad focus when searching for reflexions during the initial stages of orienting an unknown crystal. An adjustment must be provided to bring the pivot for varying the take-off angle on to a line through the X-ray tube focus.

3.6. Crystal mounting

The crystal must be exactly at the optical centre of the diffracto-meter and so it is necessary to provide three orthogonal translations

PLATE II

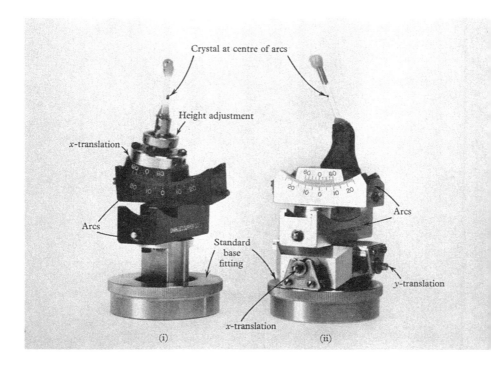

Goniometer heads: (i) eucentric, (ii) normal.

on the final crystal shaft to bring the crystal to this position. In addition, it is desirable to be able to rotate the crystal about two axes at right-angles to the crystal shaft. On a four-circle diffracto-meter sufficient rotations are already provided by the main circles, and their duplication by means of goniometer-head arcs is not essential. Nevertheless, it is often convenient to bring the crystal axes into a standard orientation relative to the diffractometer axes, and so the crystal is usually mounted on a goniometer head which has two orthogonal arcs.

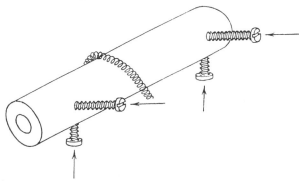

Fig. 35. Collimator fine adjustments.

Ideally, the translations, including the height or axial adjust-ment, should be above the arcs; once the crystal is brought to the common centre of the arcs, rotation about the arcs will not then demand any recentring of the crystal. Goniometer heads of this type were introduced by Furnas & Harker (1955), who termed them eucentric goniometer heads (Plate II a). They are extremely useful when the stationary crystal method (p. 21) is used for intensity measurement, as this method requires very accurate crystal centring. With moving-crystal methods this requirement can be relaxed slightly; it may then be more convenient to employ a normal kind of goniometer head with cross-slides below the arcs (Plate II b). The height adjustment is then part of the main crystal shaft. The normal goniometer head has larger traverses than those which can be provided on the eucentric goniometer head.

An internationally agreed standard exists for the bases and dimensions of goniometer heads (*I.U. Crystallographic Apparatus*

Commission, 1956), so that these, whether normal or eucentric, can readily be exchanged between diffractometers and X-ray cameras.

The orientation of the crystal by means of goniometer-head arcs leaves the rotation about the goniometer-head axis arbitrary. It should thus be possible to rotate the head about this axis relative to the zero datum point for this shaft: this datum point can then be made to coincide with a convenient standard direction in the crystal.

It may not be possible to decide the sense of the axis which has been alined along the goniometer-head axis. For example, let us

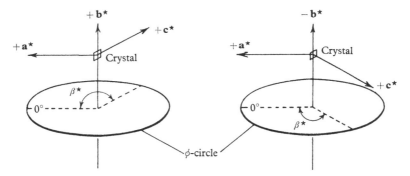

Fig. 36. Two ways of mounting monoclinic crystal with diad axis along ϕ shaft.

suppose that a monoclinic crystal is mounted about its unique axis with the **a*** direction coinciding with the zero datum point of the ϕ shaft. The unique axis may be pointing into or away from the goniometer head (Fig. 36): consequently, the rotation through the obtuse angle β^* required to bring **c*** into coincidence with the zero degree mark may be clockwise or anti-clockwise. A pre-computed list of setting angles may be usable only if the convention defining the positive sense of rotation about the goniometer-head axis can be reversed by switching: this reversal is equivalent to a change from a right- to a left-handed co-ordinate system.

A pre-requisite for rapid alinement of the crystal is a telescope which points accurately at the optical centre of the instrument. If the geometrical design of the diffractometer is such that this telescope must be removed during the intensity measurements, a well-designed kinematic mounting is essential so that the telescope can be exactly replaced. It is then possible to check whether the crystal has moved during an experimental run.

3.7. The provision of special environments

Many X-ray and neutron studies are carried out at temperatures other than room temperature; some neutron studies demand a strong magnetic field at the sample. Crystals, furnaces and magnets

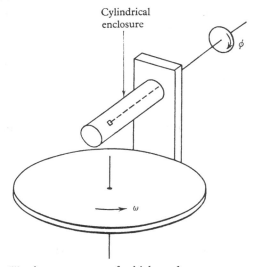

Fig. 37. Fixed-χ arrangement for high- or low-temperature work.

all require space near the specimen and the provision of these facilities may decide the geometrical arrangement adopted for the diffractometer. The fitting of the attachments tends to be difficult in a standard four-circle diffractometer in which the crystal rotates about the three Eulerian axes ω, χ and ϕ. By adopting the fixed-χ geometry, in which the χ-circle is replaced by a simple bracket supporting the ϕ-shaft, it is much easier to bring ancillary equipment close to the crystal. The high- or low-temperature enclosure can then be cylindrical in shape, concentric with the ϕ-shaft; it must, of course, be capable of rotation about the ω-axis and so it is conveniently carried by the same bracket (Fig. 37). The situation is even more straightforward in the normal-beam inclination arrangement in which there is only one crystal shaft whose direction is fixed in space.

We have now described the principal parts of a diffractometer. It is difficult to discuss diffractometer performance in the abstract: we shall, therefore, describe in detail in the following sections two specific instruments with whose design and construction we have been directly concerned, and compare their performance under the headings listed on p. 65. Our selection of these two instruments arises from our experience with them and from the fact that they present a maximum contrast in geometrical arrangement and in mechanical design. This selection does not imply that we consider them superior in all respects to the many other instruments to which we can only refer in tabular form on p. 97!

3.8. The Hilger–Ferranti four-circle diffractometer

This is a digitally-set automatic diffractometer employing normal-beam equatorial geometry. There is an X-ray version (Arndt, 1963) and a neutron version (Arndt & Willis, 1963a) of this instrument. Its design is based on two earlier instruments, an X-ray diffractometer (Arndt & MacGandy, 1962) and a neutron diffractometer (Arndt & Willis, 1963b).

The basic parts of the diffractometer are common to both versions, as are the circuits which control its automatic operation. We shall here describe only the X-ray instrument. The neutron diffractometer differs from its X-ray counterpart in that it has a larger, heavier detector arm which is in part supported by an auxiliary track; also, the entire bracket which supports the X-ray tube and collimator assembly is omitted in the neutron machine.

A photograph of the diffractometer is shown in Plate III and a drawing of its principal parts in Fig. 38. The three crystal shafts or circles (ω, χ and ϕ) and the detector shaft (2θ) can be independently positioned by a moiré fringe method. Each shaft carries a radial transmission grating which moves past a fixed pick-up head; this pick-up head contains a small fixed grating, whose lines are at a small angle to the moving grating, and a lamp and photocell assembly. Moiré fringes are generated by the relative movement of the two gratings (Fig. 39), and the movement of the fringes past the photocells produces 'add' or 'subtract' pulses which are counted on solid-state counting circuits. The positioning accuracy is 0·01°. In addition, a reference mark is provided on each circular grating which is detected by a separate reference pick-up head; this mark locates the datum point of the shaft to a reproducibility of 0·01°. Small adjustments of the χ, ω and 2θ reference pick-up heads allow the datum points to be brought exactly to zero degrees; the ϕ-grating can be rotated with respect to the goniometer head, allowing the zero value of ϕ to be brought into coincidence with any direction in the crystal normal to the ϕ-shaft. The absolute positions of the circles can also be read by means of scales, revolution counters and micrometer heads.

Each shaft is provided with a d.c. motor with an integral tacho-generator. The slewing speed is 20°/s for the 2θ shaft and 10°/s for each of the other shafts.

Rotation of the crystal through 360° is possible about the ϕ, χ and ω shafts, although the range of ω—and of χ at high values of ω—may be limited if long source or detector collimators are used. The maximum 2θ-angle of the detector arm is 155° on one side of the incident beam and 115° on the other. Limit switches are provided on all shafts.

The final crystal shaft has a fitting to accept a standard goniometer head. It has an axial adjustment of 2 cm.

The front surface of the main casting carries a bracket supporting a horizontal plate on which the shock-proof shield for the X-ray tube is mounted. This shield is provided with fine adjustments for small axial and vertical translations of the X-ray tube, for a small rotation about the horizontal axis of the tube to bring the line focus into the horizontal plane, and for a rotation about a vertical axis through the target face to allow a variation in take-off angle between $\frac{1}{2}°$ and 5°. A series of solenoid-operated filters and a fail-safe lead shutter are contained in the tube housing. Interchangeable source collimators fit into a labyrinth which forms part of the tube shield. A beam stop is mounted on a clip fitting on the collimator: it can only be used when full 360° rotation about ω is not required.

The detector arm has provision for a side-window xenon-filled proportional counter for the softer X-radiations or for an end-window scintillation counter for the detection of harder X-rays. A solenoid-operated balanced-filter assembly is sited between the

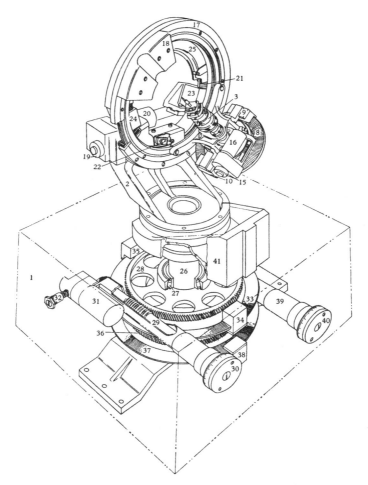

Fig. 38. Four-circle diffractometer. For key see legend of Plate III facing this page. (Courtesy Hilger and Watts Ltd.)

detector-collimator and the detector. The top or bottom and the right or left halves of this collimator can be obscured by solenoid-operated alinement shutters.

PLATE III

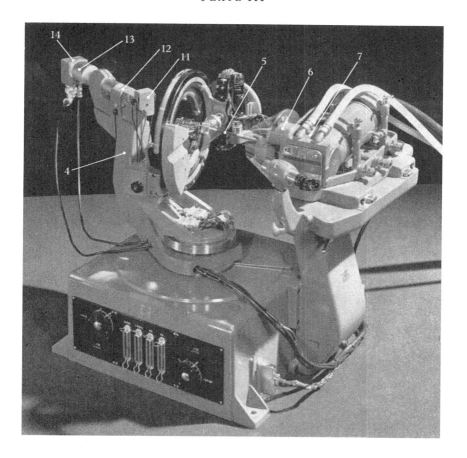

Four-circle diffractometer. The following key applies to Plate III and to Fig. 38 (opposite). 1, Diffractometer base; 2, ϕ/χ bracket; 3, goniometer head; 4, detector arm; 5, arm carrying beam stop; 6, filter and shutter assembly; 7, X-ray tube; 8, ϕ-grating; 9, ϕ main pick-up head; 10, ϕ reference pick-up head; 11, top/bottom and left/right mask assembly; 12, balanced filter assembly; 13, scintillation counter; 14, pre-amplifier; 15, ϕ micrometer dial; 16, ϕ motor and tachogenerator; 17, χ-circle; 18, counterweight; 19, χ micrometer dial; 20, χ motor and tachogenerator; 21, χ grating; 22, χ limit switch; 23, χ main pick-up head; 24, χ reference pick-up head; 25, protective cover for χ-grating; 26, hollow 2θ shaft; 27, tapered roller race; 28, 2θ worm wheel; 29, 2θ worm; 30, 2θ micrometer dial; 31, 2θ motor and tachogenerator; 32, spring loading for 2θ worm; 33, 2θ grating; 34, 2θ main pick-up head; 35, 2θ reference pick-up head; 36, ω worm wheel; 37, ω grating; 38, ω main pick-up head; 39, ω worm housing; 40, ω micrometer dial; 41, bracket for detector arm (courtesy Hilger and Watts Ltd).

PLATE IV

The Royal Institution Linear Diffractometer. A, B, C, slides; G, crystal; D, X-ray tube; P, carriage; T, oscillation mechanism; X, counter; Y, space for balanced filters (courtesy Hilger and Watts Ltd).

An incremental shaft-setting method is adopted, so that a signed increment is specified for each shaft in turn, and the arrival at the correct position is determined by a count-down to zero. The shafts

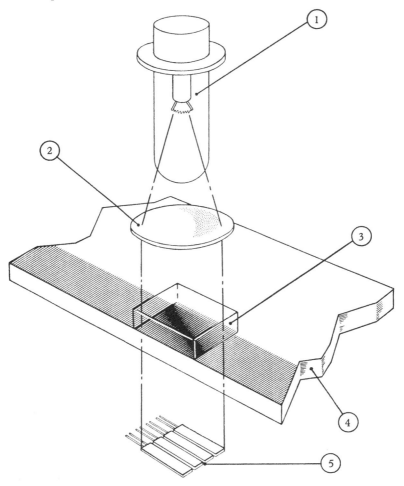

Fig. 39. The moiré-fringe digitizing system. Key: 1, exciter lamp; 2, collimating lens; 3, index grating; 4, scale grating; 5, photocell strips. (Courtesy of Ferranti Ltd.)

are set one at a time and a single-increment register is used. The slewing speed is reduced to a creep when the shaft is within half a degree of its final position.

Clamping of the shafts after setting is achieved by providing auxiliary registers with a capacity of $\pm 0.05°$ which remain coupled to their setting-motor servo-circuits during the disengagement of these circuits from the main incremental register. A small departure of a circle from its desired position through vibration or gravity is thus corrected by its own live servo-circuit.

The setting instruction for a particular shaft is in the form of an order with the following constituents:

An initiating symbol.

A function code, consisting of two decimal digits, which indicates that this is a shaft-setting instruction.

The address of the shaft to be set.

The signed incremental rotation.

A terminating symbol.

Other instructions govern the making of measurements and initiate auxiliary operations such as the insertion of balanced filters and of attenuating foils. The complete order code, together with a few special characters, is listed in Table V. The execution of one instruction leads inevitably to the reading-in of the next order, which alone can initiate the next operation.

The instructions are punched on paper tape, and pass through a reader into decoding circuits. All input information is monitored by a teleprinter; for testing purposes, its keyboard can be used as an alternative input device.

The clock pulses for the solid-state shift register, which acts as a routing circuit and distributes the decoded signals to appropriate stores, are taken from the synchronizing contacts of the teleprinter.

The detector pulses, after amplification and discrimination, are counted in a scaler. After each measurement the accumulated count is recorded by the teleprinter, both in plain language and on paper tape. The scaler is gated by signals from the main control unit; closing of the counting gate initiates a print-out followed by resetting of the scaler.

Clock pulses for the timing of individual counts are derived either from a 20 kc/s crystal oscillator or from a second detector which monitors the incident beam via a scaler with a scaling ratio

TABLE V. *The order code of the four-circle diffractometer*

Order	Function	Address	Contents
x 01	Set shaft position	Any shaft	± 5 decimal-digit-number of 1/100° units
x 02	Set ω-shaft and simultaneously set 2θ-shaft at twice the speed of ω	ω-shaft	± 5 decimal-digit-number of 1/100° units
x 03	Set filter or shutter solenoid	Any 1 of 5 solenoids	—
x 04	Reset filter or shutter solenoid	Any 1 of 5 solenoids	—
x 05	Set shaft by specified amount to bring it back to datum; print error (±); set fault register if error exceeds ±4/100°; disengage instrument after two faults unless fault count override key is on	Any shaft	5 decimal-digit-number of 1/100° units
x 06	Disengage instrument	—	
x 11	Carry out measuring scan with one shaft moving in steps, each step timed by output of variable-ratio monitoring scaler; at end of scan, print-out and reset measuring scaler	Any shaft	± or *; 2 digits (plus necessary number of significant zeros) specifying monitor counter ratio; 3 digits specifying number of steps; 1 digit p specifying the step size $2^p/100°$, where $p = 0$ to 6. The direction of the scan is specified by the sign ±; if the sign digit is * (in 5-hole code) measurement is made without movement
x 12	As for x 11, but with 2θ moving through steps of twice the size of ω-steps	ω-shaft	As for x 11
x 21	As for x 11, but with print-out and scaler reset after each step	Any shaft	As for x 11
x 22	As for x 21, but with 2θ moving through steps of twice the size of ω-steps	ω-shaft	As for x 11

Letter shift: after this character is read, all succeeding characters are ignored until the next 'figure-shift' is received. If 'letter-shift' occurs in the middle of an order the preceding codes are erased. (This is used in correcting orders set up on the keyboard.)

> character: if a limit-switch on any circle is operated during the program, the tape is searched until the next > character is found.

> character normally precedes x 05, the order for returning to datum, and appears with x 05 at regular intervals in the program tape.

< character: this provides a conditional disengage signal when the stop/run key is set at stop. It is used in program testing.

of up to 10^5. The scaling ratio is under program control and is specified in the measuring orders.

The control circuits of this system are so designed that they can set up to sixteen shafts. They can, therefore, control several diffractometers in an interleaved fashion: as soon as the first instrument is set and commences a measurement, the second diffractometer is set, and so on. For this type of operation each diffractometer has its own input and output channels. The individual diffractometers need not be mechanically identical: it is, for example, possible to operate two four-circle diffractometers, a single-axis powder diffractometer and a triple-axis spectrometer from the same control circuits. The sole requirement is that the setting operation for every instrument must be much more rapid than the measuring operation, in order to prevent one instrument having to wait for another: the individual diffractometers can make their measurements simultaneously, but only one shaft can be set at a time.

Instead of receiving its input on punched-paper tape and producing its output on the same medium, the installation can be connected directly to a computer; information transfer can then take place at a much faster rate. Two four-circle X-ray diffractometers are being controlled in this fashion in the laboratory of one of the authors by means of a time-sharing process-control computer linked to the diffractometer circuits. The advantages of on-line control are discussed in Chapter 12.

3.9. The Royal Institution linear diffractometer

This is an inclination instrument which is set by analogue methods. The principles of its design have already been described in §3.3, and a full description of the diffractometer has been given by Arndt & Phillips (1961). Its adaptation to the measurement of several reflexions at a time has been discussed by Arndt, North & Phillips (1964).

The main features of the design can be seen in Plate IV. The X-ray tube D on an adjustable mounting, the input collimator and an alinement telescope are carried on a bracket attached to the main frame of the instrument. The crystal on a standard goniometer head is mounted at G. Either a scintillation or a proportional counter can be fitted at X, and a balanced filter assembly can slide into the space Y. The scale of the instrument is dictated by the length on the slides representing one reciprocal lattice unit: in the present diffractometer this length is 5 in., giving an overall width to the instrument of 30 in.

The position of the carriage P on the three slides is controlled by means of lead screws, all of which are cut with 20 threads/in.: the counters which indicate revolutions and fractions of a revolution of the lead screws thus read directly in decimal divisions of reciprocal-lattice units. The screws in slides A and B are rotated by means of pulse-driven motors.

The slides A and B extend to ± 1 reciprocal lattice unit while slide C extends only from zero to $+1$ r.l.u.: the instrument is thus capable of measuring those reflexions in a hemisphere of reciprocal space for which $-60° \leqslant \Upsilon \leqslant +60°$ and $0 \leqslant \mu \leqslant 30°$.

The slide system is used merely to set the crystal and detector to the appropriate positions for each reflexion in turn. The carriage P is moved in a series of equal steps on the scanning slide, where each step corresponds to a unit reciprocal lattice translation. At the end of a step, the motor which moves the carriage stops, and control is assumed by the oscillation mechanism T. This device (Arndt, Faulkner & Phillips, 1960) rotates with the crystal during the motion along the slide; its own motor then imparts an additional oscillation to the crystal. The oscillation range 2Δ (Fig. 40) can be varied between $\frac{1}{2}°$ and $5°$ by means of a simple manual adjustment. During the measuring cycle the three quantities n_1, n_2 and N which represent the two background counts and the count in scanning across the reflexion are recorded both in plain language and on punched-paper tape.

After the measurement the scanning slide is moved by a fixed step to the next reflecting position, and so on, until a limit switch on the detector arm is reached. When this switch is activated, the translation currently taking place on the slide is completed (otherwise the instrument would be mis-set for the next row), the intensity of the last reflexion in the row is measured and recorded, and the carriage P is then moved one step on the stepping slide to the next parallel row. Again this row is scanned in the same way, though in the opposite direction, and the sequence is repeated until one-half of the level has been scanned and a limit switch on the stepping slide is reached.

The limit switches on the detector arm are movable: they can be set so that all reflexions within a chosen range of Υ (2θ in the

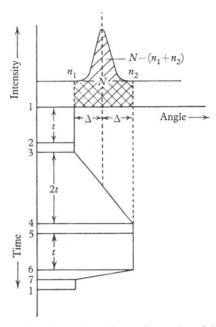

Fig. 40. Variation of X-ray intensity with angular setting of the crystal, together with the motion of the crystal during one measuring cycle and the times of signals required by the counting circuit. Key: 1, switch on scaler to measure the background on one side of the Bragg reflexion; 2, print out this count n_1 and re-set the scaler; 3, switch on scaler to measure the integrated count as the crystal sweeps through the reflecting position; 4, print out this count N and re-set the scaler; 5, switch on scaler to measure the background on the other side of the Bragg reflexion; 6, print out this count n_2 and re-set the scaler as the crystal turns back through the angle 2Δ; 7, switch off the oscillation motor and switch on appropriate shaft-setting motors to set the crystal and X-ray detector for the next reflexion (after Arndt, Faulkner & Phillips, 1960).

zero level) are measured. Additional safety switches are fitted at the ends of the slides.

The sequence of operations is summarized in Fig. 41, from which it will be seen that two facilities have not yet been mentioned. A count is kept of the indices of the reflexions and these indices are recorded along with the corresponding intensity measurements.

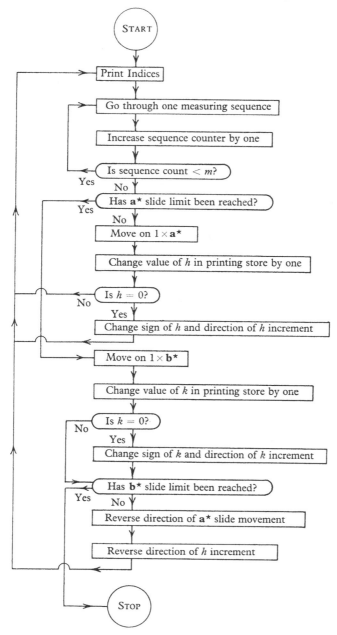

Fig. 41. Sequence of operations of linear diffractometer. The sequence counter records the number of oscillation cycles; it can be preset to any number *m* (after Arndt & Phillips, 1961).

Also each reflexion can be measured more than once: the number of measuring cycles can be selected by a hand switch.

All the reflexions are measured in a systematic order. For example, with the scanning slide and stepping slide parallel respectively to **a*** and **b*** and the vertical slide parallel to **c*** the index l is the slowest-moving index, changing only when the level is changed by hand, and the index h is the fastest moving, changing by one for each step on the scanning slide. The index k changes by one for each step on the stepping slide.

The control system includes three stores for the three indices; one of these consists simply of switches which are set by hand to the appropriate value for the level which is being measured, while the other two are bi-directional uniselectors. Their settings are read and printed out at the beginning of each cycle of intensity measurements. The order in which the three indices are printed can be selected by means of three key switches to suit the particular crystal under investigation. The choice of scanning slide (either A or B in Plate IV), that is, of the fastest moving index, is also made by means of a key switch.

Distances moved along the slides are measured by counting the number of electrical pulses fed to the driving motors. One pulse corresponds to a motion along a slide of 0·0001 reciprocal-lattice units. These pulses are fed to the appropriate batch counter, which can be set by means of four-decade switches to four-digit-number totals representing ten-thousandths of 1 r.l.u. When the correct number of pulses has been received, the counter automatically resets itself and is ready to accept another batch of the same size; at the same time the counter produces a signal which is used to stop the motor pulse generator and to initiate the next operation in the sequence.

Additional switching circuits permit one or the other of a pair of balanced filters (p. 182) to be inserted in alternate measuring cycles.

It is also possible to switch the sequencing circuits, so that reflexions absent because of space-group symmetry are automatically skipped.

3.10. A comparison of the four-circle and linear diffractometers

The suitability for a particular purpose of one or the other of the two diffractometers which have been described in the last two sections depends upon the relative importance which is attached to the various factors listed on p. 65.

Accuracy of setting and of intensity measurement

An important difference between the two instruments is in the realm of setting accuracy. The maximum setting error of about

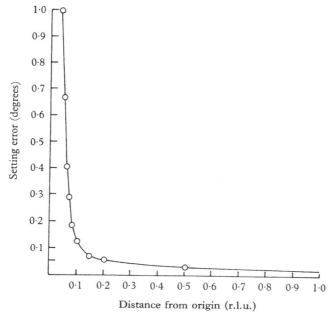

Fig. 42. Setting errors of linear diffractometer (after Arndt, 1963).

0.02° for any shaft of the four-circle diffractometer is unlikely to be a limitation for most problems. The setting precision of the linear diffractometer has been considered by Binns (1964). For the crystal shaft (ϕ-axis) this precision varies with the distance from the origin of the reciprocal lattice level which is being investigated. Fig. 42 shows the setting errors which were found in an actual instrument.

In practice, all reflexions more than o·1 r.l.u. from the origin are set automatically with adequate accuracy.

The four-circle diffractometer allows a choice of ω-scan or $\omega/2\theta$-scan and thus permits the more suitable scan to be selected for minimizing the white radiation error (p. 227). The linear diffractometer can only be used with the moving-crystal-stationary-detector method.

In the four-circle diffractometer the azimuthal orientation of the reflecting plane can be varied. By means of an azimuthal scan it is possible to make an empirical correction for absorption (p. 241); errors due to simultaneous reflexions (p. 251) can be minimized by a suitable choice of azimuthal angle.

Speed of operation

The principal factor which determines the number of reflexions which can be measured in a day is the time needed to accumulate a count of the desired statistical significance. The setting speed of both instruments is high, and differences in their setting speeds do not greatly affect their overall speed of operation.

Both instruments can be adapted for the simultaneous measurement of several reflexions (p. 56). The necessary modifications to the standard instrument are somewhat simpler for the linear diffractometer (Arndt, North & Phillips, 1964) than for the four-circle diffractometer (Arndt & Phillips, 1966). Very fast data collection is less convenient with the linear diffractometer which requires the operator's attention at the completion of each level; with a small number of reflexions per level this means that manual intervention may be necessary every few hours.

Number of accessible reflexions

The linear diffractometer has four basic limitations on the number of accessible reflexions:

(1) The maximum Bragg angle attainable is 32·5°.

(2) Reflexions lying within o·1 r.l.u. of the origin cannot be set up automatically with adequate accuracy.

(3) Reflexions whose reciprocal lattice points lie on or near the goniometer-head axis cannot be measured.

(4) Within these limitations only a hemisphere of reciprocal space is accessible.

The four-circle diffractometer permits exploration of the entire sphere of reciprocal space (with the exception of that part shadowed by the crystal mounting) up to a maximum Bragg angle of 55° using the 'cone' or 'symmetrical-A' method of data collection. Reflexions up to the instrumental maximum of 77·5° can be measured using the general setting method (§2.4).

Amount of manual intervention required

The linear diffractometer can only measure one reciprocal lattice level automatically; the operator's intervention is required to set up the instrument for the next level and to interrupt the set sequence within one level in order to measure reference reflexions. In favourable cases the four-circle diffractometer, on the other hand, can be allowed to run unattended until a complete three-dimensional set of data has been collected. The reflexions can be measured in any sequence and reference reflexions can be inserted automatically at any desired point.

Accessibility of specimen

There is little to choose between the two instruments in respect of the space available near the specimen for special attachments (furnace, cryostat, etc.). Space is very limited in both cases.

Availability of a computer

Any modern structure determination requires the availability of an electronic computer. If the number of reflexions which have to be measured is such as to warrant an automatic diffractometer at all, then a computer is virtually essential for checking and processing the experimental data. However, it is quite possible, if time on the computer is only available at intervals, to accumulate the results obtained on a linear diffractometer before processing the data.

The four-circle instrument requires the use of a computer for the production of a program tape in addition to its employment for data processing. One way of making use of the greater versatility of this instrument is to make a preliminary run covering all reflexions. As a by-product of processing the results from this run a new program tape is prepared which can specify longer measuring periods for weak reflexions or the insertion of filters for very strong reflexions. For this way of operating the instrument, daily access to a computer, albeit for quite short periods, is necessary.

Versatility

The four-circle diffractometer is much more versatile than the linear diffractometer. Its order code permits many different measuring procedures; several quantities, such as the rocking range through a reflexion, the measuring period, the choice of filter, are under program control. This very versatility, however, makes it a more difficult instrument to use, and more experience is necessary to take full advantage of its potentialities. Where only routine data collection is required, the additional features of the programmed instrument may not be needed.

Reliability

The four-circle instrument is more complicated and contains many more circuit components than does the linear diffractometer; however, by using the principles of computer design, such as solid-state circuitry, printed-circuit-board construction and testing facilities, a degree of reliability comparable with that of a small computer can be achieved. With an installation of this kind the mean time between failures of the electronic circuitry is some 500–1,000 h. In both instruments the most unreliable component is the electro-mechanical output device (teleprinter).

Cost

As would be expected from a more complicated instrument, both the capital and running costs of a four-circle diffractometer installation are greater than those of the simpler linear diffractometer. Against this may be balanced the fact that the control circuits of a digitally-set installation can be used to control more than one instrument on a time-sharing basis.

3.11. Some other designs

Some other diffractometers which have been described are listed in Table VI. The examples which are quoted have been selected either for their historical interest or because the instruments illustrate some specific features of design. Many more instruments, especially non-automatic ones, are known to have been constructed. The list does scant justice to the many neutron instruments which have been developed as general-purpose diffractometers and which have mostly been automated to a greater or lesser extent. A recent review of such instruments has been prepared by Atoji (1964).

TABLE VI. *X-ray and neutron diffractometers*

Literature reference	Manufacturer	Remarks
1. Inclination geometry		
(a) Non-automatic X-ray:		
Cochran (1950)	—	—
Clifton, Filler & McLachlan (1951)	—	Flat-cone geometry
Evans (1953)	—	—
Mathieson (1958)	—	Linear diffractometer
Buerger (1960)	Supper	Automatic version now exists
Potter (1962)	—	Linear diffractometer
(b) Non-automatic neutron:		
Lutz (1960)	M.A.N.	—
(c) Automatic X-ray:		
Bond; Benedict (1955)	—	Spiral scan of reciprocal space
Drenck, Pepinsky & Diamant (1957)	—	Electrical analogue diffractometer
Brown & Forsyth (1960)	—	Paper-tape input; digital instruments
Clastre (1960)	Secasi	Paper-tape input; digital instruments
Ladell & Cath (1963)	Philips	Linear diffractometers
Arndt & Phillips (1961)	Hilger & Watts	Linear diffractometers
Abrahams (1962)	Supper	Paper-tape input; digital instrument
2. Normal-beam equatorial geometry		
(a) Non-automatic X-ray:		
Bragg & Bragg (1913)	—	Ionization-chamber detector
Furnas & Harker (1955)	—	—
Furnas (1957)	G.E.	—

7

TABLE VI (*cont.*)

Literature reference	Manufacturer	Remarks
(b) Automatic X-ray:		
Wooster & Martin (1936)	—	Ionization-chamber detector
Wooster & Wooster (1962)	Crystal Structures	
Arndt (1963)	Hilger & Watts	Paper-tape input; digital instruments
Mayer (1964)	Siemens	
—	Enraf-Nonius	
—	Picker	Punched-card input; digital instruments
—	G.E.	
(c) Automatic neutron:		
Prince & Abrahams (1959)	—	
Levy, Agron & Busing (1963)	—	Paper-tape input; digital instruments
Arndt & Willis (1963 a, b)	Ferranti	

3. On-line computer-controlled diffractometers

Cole, Okaya & Chambers (1963)	IBM	X-ray diffractometer controlled by IBM 1620 computer
Bowden, Edwards & Mills (1963)	—	X-ray diffractometer controlled by ICT Atlas computer
Hamilton (1964)	—	Group of neutron diffractometers controlled by SDS 920 computer
Arndt, Gossling & Mallett (1966)	—	Group of X-ray diffractometers controlled by Ferranti Argus 304

CHAPTER 4

DETECTORS

4.1. Introduction

All radiation detectors used in diffractometry function in a basically similar fashion: the detection of an individual incident X-ray quantum or neutron results in the collection of a certain quantity of electrical charge at the input terminal of the detecting circuitry. No matter what the actual detector may be, this resultant charge may be dealt with in one of two ways. Either the charge which corresponds to the arrival of a given number of pulses per unit time is integrated and the resultant current is taken as a measure of the incident intensity, or the individual pulses of charge are counted, if necessary after amplification and shaping.

The first method, that of current measurement, is used only with X-ray ionization chambers which are rarely employed today. All other X-ray and neutron detectors are 'counters' which produce discrete pulses. These detectors have the important feature that the result of an intensity measurement is given directly in digital form as the number of incident quanta or neutrons in a given time interval: such digital data are thus already in a suitable form for further processing.

It is easy to specify the requirements to be met by an ideal detector for use in diffraction studies. Such a detector should have the following properties:

(1) It should have an efficiency of unity for the radiation to be detected and zero efficiency for all other radiations, including X-rays or neutrons of wavelengths closely bordering on the monochromatic radiation employed.

(2) Its output pulses should be large compared with any noise or interference pulses which may be present; only simple low-gain amplifying equipment, if any, should be necessary.

(3) The device must be stable: its efficiency, and the amplitude and shape of the output-pulse, must not vary with time (fatigue effects), temperature or counting rate.

(4) It must have a short resolving time so that incident X-ray quanta or neutrons are counted with minimal counting losses even at high incident fluxes.

(5) The sensitive area of the device must be large enough to accept the full spread of the diffracted beam from the crystal under investigation. The sensitivity should be constant over the whole of this area.

(6) The weight and the physical dimensions of the detector, together with any necessary shielding against stray radiation and electrical interference, should be as small as possible so as to simplify the mechanical design of the detector arm.

Actual detectors fall short of some or all of these requirements, and it is necessary to examine the different devices in detail in order to compare their properties.

The production of a quantity of charge implies an initial ionizing event. Neither X-ray quanta nor thermal neutrons are efficient producers of primary ionization but both can produce secondary ionization. In a gas this ionization process consists of the formation of free electrons or negative ions and of positive ions; in a semi-conducting solid, the process consists of the production of electrons and holes which can move in opposite directions under the influence of an electric field. Thus we can distinguish gas ionization counters and solid-state ionization counters.

Another type of detector, the scintillation counter, makes use of a different principle: the energy of the incident X-ray quantum, or the energy liberated in a nuclear reaction which absorbs the incident neutron, is converted into the energy of light photons in a phosphor or scintillator. These photons, in turn, are made to produce photoelectrons at the cathode of a photomultiplier whose output is, once again, an electrical pulse.

We shall first discuss X-ray counters and use them to illustrate the general properties of these three classes of counters. We shall then deal in somewhat less detail with points specific to neutron detectors. At the end of the chapter we shall speculate briefly on the form which detectors of the future are likely to take.

4.2. X-ray gas ionization counters

An ionization counter consists of a gas-filled chamber which contains two electrodes across which a potential difference is applied. In the most common geometrical arrangement, the electrodes are two concentric cylinders; the anode is a thin central wire and the outer wall of the device serves as the cathode. This general arrangement is common to ionization chambers, proportional counters and Geiger or Geiger–Müller counters. The chief difference between these three types lies in the magnitude of the potential difference between the electrodes.

The chamber is filled with a gas at a pressure which for the longer X-ray wavelengths ($\lambda > 1$ Å) is generally about atmospheric. This gas should have a high specific ionization and a high absorption for the X-rays to be detected. In addition, for reasons which will become apparent, there must be no tendency to form negative ions. These conditions restrict the choice of filling gases to the noble gases argon, krypton and xenon, which are usually mixed with a small amount of a polyatomic vapour. It is worth mentioning at the outset that, in contrast with the situation when charged particles are being detected, only those quanta which are absorbed and destroyed within the chamber can be detected. (The same is true of neutron detection.) The incident X-radiation enters the counter through a thin window, generally of beryllium or mica. In Geiger counters this window is usually at one end of the cylindrical counter so that the path of the radiation is axial (Fig. 43a); in proportional counters, for reasons which are discussed below (p. 105), the window is at the side of the counter so that the direction of the rays is normal to the axis (Fig. 43b).

The ionization process

The X-ray wavelengths used in single crystal investigations are those of the characteristic K emission lines of the elements between chromium and silver in the periodic table. These, together with the quantum energies in kilo-electron volts, are given in Table VII. The most commonly employed target materials are copper and molybdenum.

When an X-ray quantum within this energy range is absorbed in a column of noble gas, by far the most likely event is the emission of a K-electron. Either the gas-ion so produced can then emit its own characteristic X-radiation, or internal conversion takes place which will result in the ejection of one or more Auger-electrons. With argon and xenon the Auger yield is almost complete so that

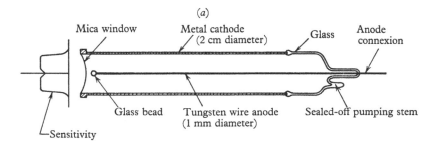

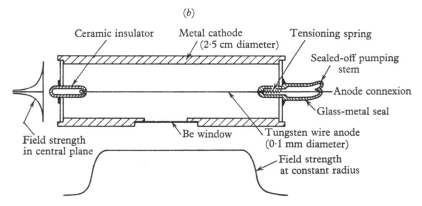

Fig. 43. (a) X-ray Geiger counter, showing sensitivity across the window. (b) X-ray proportional counter, showing variation of field strength along its length and across its diameter.

the total energy of the incident quantum becomes available as kinetic energy of the photo- and Auger-electrons. This kinetic energy is then dissipated by the production of further ionization as these electrons collide with neutral gas molecules. For example, the energy of a copper $K\alpha$ quantum is 8×10^3 eV, and the energy required to produce an ion-pair in argon is 29 eV, so that about 275 ion-pairs are produced when one quantum is absorbed in argon.

TABLE VII. *X-ray wavelengths and energies*

Radiation	λ (Å)*	E (keV)
CrKα	2·291	5·41
MnKα	2·103	5·90
FeKα	1·937	6·40
CoKα	1·790	6·93
NiKα	1·659	7·47
CuKα	1·542	8·04
MoKα	0·711	17·44
AgKα	0·561	22·11

* Weighted mean of Kα_1 and Kα_2.

There are some relatively rarer absorption events in which no internal conversion takes place and the argon characteristic X-radiation escapes from the detector; the amount of energy available for producing secondary ion-pairs is then the energy of the incident quantum less the binding energy of the argon photo-electron, which is 3·2 keV. With an incident quantum of CuKα radiation the number of ion-pairs produced, therefore, is only $\frac{1}{29}(8\cdot0-3\cdot2) \times 10^3$, or about 165.

The number of ion-pairs produced directly by the incident X-ray quanta by other processes is negligible compared with that produced by photo-electric processes.

Pulse size and dependence on applied voltage

In the absence of an electric field in the counting chamber the ion-pairs would recombine, but under the influence of a field the electrons drift towards the central anode and the positive ions are collected on the outer cathode. If the field is large enough to prevent recombination, but too small for the electrons to acquire sufficient kinetic energy to produce further ionization by collision as they drift to the anode, this is the only process and the device functions as an ionization chamber.

If, however, while traversing one mean free path, the electrons are sufficiently accelerated, further ion-pairs will be formed at each collision. This process is, essentially, the formation of a Townsend avalanche and is known as 'gas multiplication' or 'gas amplification'. The number of ion-pairs of each polarity collected is An, where A is the gas amplification factor and n is the number of primary ion-pairs.

Consequently, if the detector is coupled to a detecting circuit with time constants which are long compared with the collection times of the positive ions, the size of the voltage pulse produced at the positive collecting electrode is

$$V = -\frac{Ane}{C}, \qquad (4.1)$$

where $e = 1 \cdot 6 \times 10^{-19}$ coulombs is the electronic charge and C is the capacity of the counting chamber and of the connected circuit components. Equation (4.1) gives the final amplitude of the pulse at the instant of positive-ion collection. The rise time of the pulse and its shape are discussed on p. 110. In a typical case C might be $30\mu\mu F$; for a CuKα quantum and an argon-filled counter, $n = 275$, and so

$$V = 1 \cdot 45A \times 10^{-6} \text{ V}.$$

In practice, the amplifier time constant is not long compared with the ion collection time, and the actual pulse is several times smaller than that indicated by equation (4.1). It is impossible to design an amplifier in which the amplitude of noise pulses, referred to the input stage, is less than about 10 μV and so the gas amplification factor must be at least 50 to obtain an adequate signal-to-noise ratio. Individual CuKα quanta cannot be counted by means of an ionization chamber for which $A = 1$. Generally values of $A \simeq 10^3$ are used, necessitating electronic amplifier gains of between 10^3 and 10^5.

In the range from 10 to 10^4, A varies approximately with the exponential of the applied voltage; for a typical X-ray counter an increase of 100 V in the applied voltage leads approximately to a trebling of the value of A. In the lower part of the range the gas multiplication is confined to a region very near the central anode, but as the voltage increases this region extends over an increasing part of the counter. The discharge still remains confined, however, to the immediate vicinity of the plane, normal to the counter axis, in which the original ionization process took place. While the avalanche remains localized, the gas amplification A is independent of the number of primary ion-pairs n and the size of the final pulse (equation 4.1) is proportional to n and hence to the energy of the incident X-ray quantum. A device functioning in this way is called a 'proportional counter'. The exact value of A at a given voltage is strongly dependent on the purity of the noble-gas filling: in order to avoid changes in A as the counter ages, and adsorbed impurities are given off by the walls of the counter, it is usual to add a few per cent of a polyatomic stabilizing substance to the filling. CO_2 and C_2H_4 are commonly used for this purpose.

In conjunction with suitable pulse-height analysing circuits, proportional counters may be used for discriminating between X-rays of different wavelengths. Obviously it is necessary for the gas amplification to be independent of the region in the counter in which the initial ionization occurs. Thus X-ray proportional counters are nearly always of cylindrical geometry, as the radial distribution of the field strength is such that all the gas multiplication takes place in the immediate vicinity of the central wire. The collimated X-ray beam usually enters the counter through a side-window somewhere near the central plane of the cylinder where the field is free from end-effects (Fig. 43 b). Consequently, the column of absorbing gas is necessarily short. The percentage absorption of CuKα and MoKα X-rays in the noble gases is shown in Fig. 44 as a function of column length and gas pressure. In order to keep the operating voltage of the proportional counter reasonably low, it is preferable not to fill the counter to a pressure exceeding atmospheric. Krypton has properties which make it an undesirable filling for proportional counters (see p. 108). Fig. 44 shows that xenon is preferable to argon as a filling by virtue of its higher absorption. MoKα radiation cannot be detected efficiently in a proportional counter.

As the voltage applied across the electrodes of the counter is further increased, the amplification factor continues to increase but it is now no longer independent of n. When A reaches 10^8 or 10^9 all pulses produced in the counter are of the same size. At this stage the discharge travels along the whole length of the counter; it is propagated by ultra-violet photons formed in the original avalanche. The positive ions, as they reach the cathode, can produce further avalanches. The counter now behaves as a Geiger counter; in the absence of 'quenching' the discharge becomes continuous.

The variation of pulse size with applied voltage for CuKα and MoKα radiations is shown in Fig. 45.

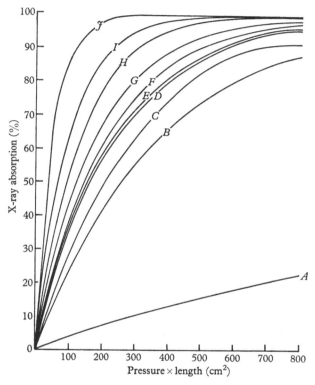

Fig. 44. X-ray absorption as a function of the product of gas pressure in cm of mercury and column length in cm. *A*, MoKα in A; *B*, CuKα in A; *C*, MoKα in Xe; *D*, CoKα in A; *E*, MoKα in Kr; *F*, CuKα in Kr; *G*, FeKα in A; *H*, CoKα in Kr; *I*, FeKα in Kr; *J*, CuKα in Xe (Arndt, 1955).

Pulse-height distribution and pulse-height analysis

The formation of a Townsend avalanche is a statistical process: the number of final ion-electron pairs formed, and thus the amplitude of the output pulse, are subject to statistical fluctuations. For this reason, when a monochromatic beam of X-rays is detected by a proportional counter, the pulse-height distribution curve is not a sharp line but has an approximately Gaussian distribution about the mean. The variance σ_n^2 of the distribution is equal to kn, where n is the original number of ion-pairs formed; the factor of proportionality k is approximately unity, but its exact value cannot be predicted in an entirely straightforward way (see, for example, Wilkinson, 1950, who reviews some of the work on this subject). The fractional standard deviation is

$$\frac{\sigma_n}{n} = \sqrt{\frac{k}{n}} \simeq n^{-\frac{1}{2}}.$$

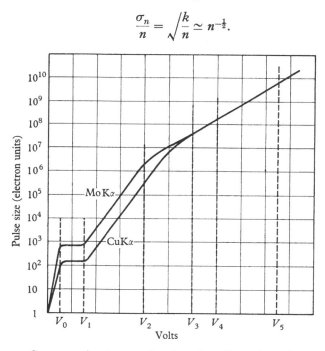

Fig. 45. Counter pulse size as a function of applied voltage: below V_0, partial recombination; V_0 to V_1 ionization chamber region, $A = 1$; V_1 to V_2, proportional region, A independent of energy; V_2 to V_3, region of limited proportionality; V_4, Geiger threshold voltage; above V_5, continuous discharge (Arndt, 1955).

Fig. 46 shows the number of pulses of amplitude between V and $V + \delta V$ plotted against V, the amplitude of the pulses, for a xenon-filled proportional counter receiving a crystal-reflected beam of CuKα radiation. Both the energy and wavelength scales corresponding to this arbitrary amplitude scale are shown. The main peak is due to CuKα radiation; it has a full width at half height of

15 per cent of the mean energy. The small peak at 4 keV, called the 'escape peak', is caused by the small proportion of absorption events in which energy is lost in the form of xenon L radiation which escapes from the counter. The peak at 16 keV is produced by the first harmonic of wavelength 0·77 Å reflected from the crystal monochromator, and the rise of the counting rate at very small pulse heights is the result of amplifier noise.

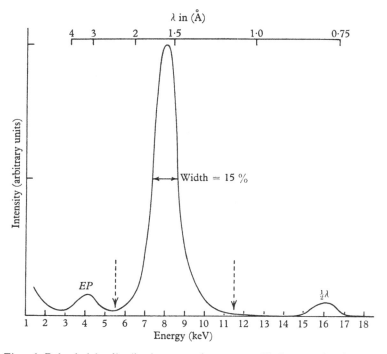

Fig. 46. Pulse-height distribution curve for a xenon-filled proportional counter with crystal-reflected Cu Kα radiation, showing the escape peak *EP* and the $\frac{1}{2}\lambda$ component. The normal setting of the 'window' is indicated by the arrows.

It is worth noting here that the amplitude of the escape peak is not always as small as in Fig. 46. The yield of KrKα radiation, when krypton is irradiated with MoKα radiation, is very high and the pulse-height distribution curve of a krypton-filled counter shows an escape peak whose amplitude is greater than that of the main peak (Fig. 47). For this reason krypton-filled proportional counters are rarely used.

Curves such as those of Figs. 46 and 47 are obtained with an electronic circuit variously known as a single-channel pulse-height analyser, pulse-height discriminator, or 'kicksorter' (p. 158). Such a circuit will accept and pass on to the electronic counter or counting-rate meter only pulses which lie within its 'window' δV, or those which have amplitudes between voltages V and $V + \delta V$. Fig. 46 was obtained by keeping the window width or channel

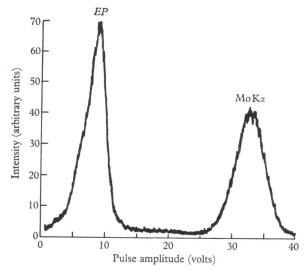

Fig. 47. Pulse-height distribution for a krypton-filled proportional counter with MoKα radiation, showing large escape peak (EP) (after Parrish & Kohler, 1956).

width δV constant at 0·5 V and varying the value of V. In normal use, for X-ray intensity measurements, the window is made much wider; it is usual to set it symmetrically about the peak of the distribution and to make its width such that about 90 per cent of the main peak is accepted and only the extremes of the tails of the peak are rejected. A sharper discrimination not only causes a loss of overall sensitivity of the system, but also makes great demands on the stability of the amplifier and of the high-tension supply.

In X-ray diffraction studies there are three functions in which wavelength discriminating properties of a detector are beneficial. They are the separation of crystal-reflected characteristic radiation from its harmonics, the discrimination against general white radiation, and the separation of the characteristic radiation from any fluorescent radiation which may be excited in the specimen. When a pulse-height discriminator is used, the complete suppression of harmonics is easy, the attenuation of white radiation is useful but not complete, and the degree of attenuation of fluorescent radiation is only just worthwhile. (In X-ray fluorescence analysis a very narrow analyser window may be used specifically in order to distinguish between the characteristic radiations from neighbouring elements: this is a different problem from that discussed here.) Discrimination between the Kα and Kβ components of the characteristic radiation is achieved only to a barely perceptible extent. The use of pulse-height discrimination, either alone, or in conjunction with other monochromatizing techniques, is discussed further on p. 180.

Proportional and Geiger counter pulse shapes

We have already stated that the full amplitude $V = -Ane/C$ of the proportional counter pulse is reached only at the instant when the positive ions are collected. The form of the function of voltage with time, known as the pulse shape, will now be considered in more detail, following the treatment given by Wilkinson (1950).

At any instant of time t, the potential at the positive collecting electrode will be

$$V(t) = \frac{q_+(t) + q_-(t)}{C}, \qquad (4.2)$$

where $-q_+$ is the charge induced by the positive charges and $-q_-$ that induced by the negative charges. In the counters which are of interest in our discussion the presence of electronegative filling gases, such as oxygen, is carefully avoided in order to prevent the formation of slow-moving heavy negative ions. The negative charges are unattached electrons which are collected in a time t_1; t_1 is very small compared with t_2, the collection time of the positive charges (positive ions).

At the instant of collection of the electrons, $q_- = -Ane$ and

$$V(t_1) = -\frac{Ane}{C} + \frac{q_+(t_1)}{C}, \qquad (4.3)$$

and at the instant of positive ion collection $q_+ = 0$ and

$$V(t_2) = -\frac{Ane}{C},$$

in accordance with equation (4.1). The calculation of $V(t)$ requires a knowledge of $q_+(t)$ and in the general case this latter function depends on the location within the counter of the point of origin of the positive ions. $V(t)$ can thus vary from pulse to pulse. In a proportional counter (and also in a Geiger counter) of concentric-cylinder geometry, however, the distribution of electric field strength within the counter is such that the overwhelming majority of secondary ion pairs originate within a few wire diameters of the anode, and consequently there is no spatial variation of $V(t)$ as there is with ionization chambers.

It follows that with proportional counters $q_+ \simeq Ane$ when t is so small that the positive ions have not yet moved far from the central wire, thus at the instant of electron collection $V(t_1)$ is still very small. The major part of the pulse is due to the motion of the positive ions away from the immediate vicinity of the central wire; equation (4.2) shows that the pulse only reaches its final height at the moment of positive ion collection ($q_+ = 0$). The complete pulse shape can be calculated from a knowledge of the field distribution and the positive-ion mobilities, and it can be shown that the potential reaches one half of its final amplitude at the instant $t_{\frac{1}{2}} = (a/b) \times t_2$, where a is the wire radius and b the radius of the cathode of the counter. In a typical X-ray counter, t_2 is a few hundred microseconds and $t_{\frac{1}{2}}$ is a few microseconds. The final pulse shape is as shown in Fig. 48, calculated for a typical counter in which $b/a = 100$.

Pulse shaping

In the above discussion it has been assumed that the time constants of the circuits coupled to the collecting electrode are long compared with the ion-collection times. This is undesirable in practice, since the superposition of pulses such as those of Fig. 48 leads to a d.c. build-up which can bias off the amplifier connected to the detector. In any practical amplifier system the coupling between the different stages of the amplifier introduces a 'differentiation' of the pulse shape. The amplifier may contain a number of networks of the type shown in Fig. 49 *a* but it is customary to make the time constant of one of these networks very much shorter—say 100 times—than the time constants of all the others. The differentiation of the original pulse shape can then be regarded as being due solely to this one short time constant. The passage of the pulse of the shape shown in Fig. 48 through such a network, sometimes

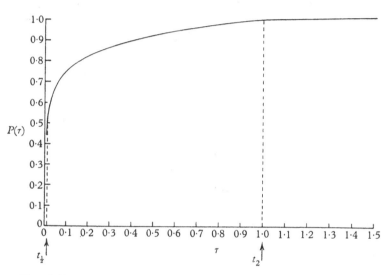

Fig. 48. Proportional-counter pulse height before differentiation as a function of $\tau = \text{time}/t_2$, where t_2 is the collection time for the positive ions. $P(\tau)$ is expressed as a ratio of the final pulse height (after Wilkinson, 1950.)

called a 'clipping network', produces the time derivative of the input pulse: the effect of varying the sharpness of differentiation is shown in Fig. 50. The effect of reducing the clipping time constant is to reduce the response of the amplifier at low frequencies.

As Fig. 50 shows, quite sharp differentiation is possible with proportional-counter pulses without excessively reducing the size of the pulse: microsecond pulses are produced, even though they originate from slow-moving positive ions. It is usual to follow the differentiating network with an integrating network of the type shown in Fig. 49 *b*. A network of this type lengthens the trailing edge of the pulse and reduces the response of the amplifier at high

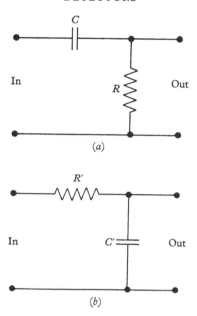

Fig. 49. (a) Differentiating network; (b) integrating network.

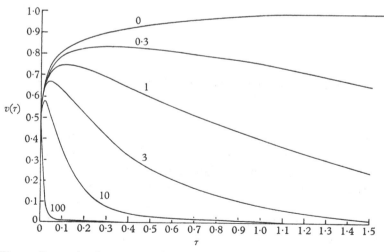

Fig. 50. Proportional-counter pulse shape after differentiation of varying sharpness s. s is the ratio of the positive ion collection time t_2 to the clipping time constant RC. τ is the time after the start of the pulse, expressed as a fraction of t_2. v is expressed as a ratio of the final pulse height before differentiation ($s = 0$). s values are shown on the curves (Wilkinson, 1950).

8

frequencies and hence its response to certain types of noise generated in the amplifier. For most purposes the optimum signal-to-noise ratio can be secured if the differentiating time constant, $\tau_{\text{diff.}} = RC$, and the integrating time constant, $\tau_{\text{int.}} = R'C'$, are equal (Gillespie, 1953). However, the high-frequency response of any practical amplifier is limited even without a specific integrating network (for example, by the cut-off frequency of the transistors employed) and the penalties of not optimizing the upper frequency cut-off of the amplifier are not large.

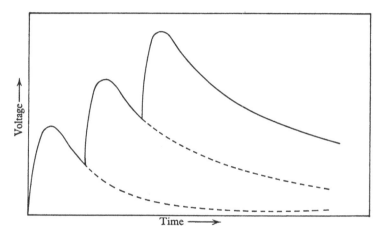

Fig. 51. The pile-up of pulses at high counting rates produces a d.c. voltage component.

Since the pile-up of pulses illustrated in Fig. 51 introduces a d.c. component, it is necessary for the differentiation to be carried out before the signal level is high enough to affect the bias voltages of the amplifier stages. Input pulses from X-ray proportional counters are about 1 millivolt in amplitude. The differentiating and integrating networks are placed at a point in the amplifying chain where the signal is 20–50 mV in amplitude, in order to avoid attenuating the input signals too early.

High counting rates in proportional counters

With sharper differentiation, that is, with differentiating networks of very short time constants, amplified pulses can be produced from a proportional counter which have a duration much less than $1 \mu s$, and in theory counting rates as high as $10^7/s$ are possible. However, a proportional counter is employed in X-ray diffractometry largely by virtue of its wavelength discriminating properties and so it is important not only that detectable pulses should be produced over the whole range of counting rates envisaged, but also that the gas multiplication factor A should be constant. There is no change in A until the spacing of the pulses becomes equal to the collection time of the positive ions, which is a few hundred microseconds for a typical X-ray counter. At higher intensities, however, the positive ions which are present in the counter from previous discharges shield the region near the central wire and lead to a reduction in field strength and hence in gas amplification. The effect is obviously larger the larger the individual avalanches. For fast counting, therefore, the initial gas amplification should be kept small (< 500). Even then, pulse-height discrimination ceases to be practicable at counting rates greater than 5×10^4 counts/s. This is the maximum practicable counting rate, and no advantage is gained by working with extremely short time constants: a value of $\tau_{\text{diff.}}$ of $1-2 \mu s$ is short enough to reduce pile-up to negligible proportions and to produce sufficiently sharp pulses without excessive attenuation.

Coupling of proportional counter to amplifier

The precise way in which a proportional counter is coupled to the amplifier affects the pulse height and can influence the differentiation of the pulse profile. Two alternative methods of coupling a proportional counter are illustrated in Fig. 52. Method (*a*) has the advantage that no blocking condenser is needed. The cathode of the counter, which is usually also its case, is at high voltage and the counter must be enclosed in, but insulated from, an earthed container. Method (*b*) requires a blocking condenser. This method was formerly avoided since any electrical leakage across the condenser gives rise to spurious voltage pulses, but modern condensers are much better in this respect and the method is now generally preferred.

The capacity of the blocking condenser is generally considerably larger than C, the capacity of the counter and its associated leads.

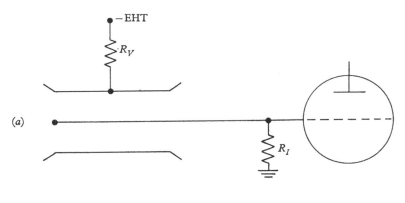

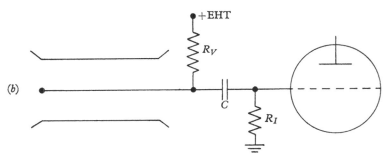

Fig. 52. Coupling of proportional counter to amplifier: (a) without and (b) with blocking condenser.

Consequently, in both methods of connecting the counter to the amplifier, the input time constant of the system is

$$\tau_{\text{in.}} = C\left(\frac{1}{R_V}+\frac{1}{R_I}\right)^{-1},$$

where R_V is the high-voltage series-resistor and R_I the input resistance of the amplifier. In order to give the maximum pulse amplitude, C must be as small as possible, since the pulse amplitude is inversely proportional to C; typically, as we have seen, C will be about 30 $\mu\mu$F. τ_{in}. must be large compared with the differentiating time constant of the amplifier which is about 1 μs; accordingly, R_V and R_I should have values of several megohms. R_V can readily be made large, and a large input resistance R_I is easy to achieve with valve amplifiers and is quite possible with transistor amplifiers.

In all diffractometers the counter is mounted on some form of moving detector arm and the amplifier is usually at some distance from the detector itself. The capacity of screened cables is 10 $\mu\mu$F or more per foot, and so the total cable capacity is comparable in magnitude with, or greater than, the capacity of the detector. Thus it is very desirable to have, immediately adjacent to the counter, a pre-amplifier from which the pulses are fed into the cable via a cathode or emitter follower. The output impedances of cathode followers are low, so that they effectively prevent the attenuation of the counter pulse by the capacity of the cable, which shunts the input impedance of the main amplifier.

Performance of proportional counters

The efficiency of a typical commercial X-ray proportional counter when used with and without pulse-height discrimination is shown in Fig. 53a. The sensitivity to radiations other than X-radiation in the range 0·5–3·5 Å is very small, and under normal laboratory conditions the non-X-ray background should not exceed 10 counts/min.

X-ray proportional counters have been considered in considerable detail since a discussion of their properties and of the associated electronic circuits forms a suitable introduction to other detectors. The extent to which proportional counters fulfil the requirements of an ideal detector can be seen from Table VIII.

Proportional counters are at present the most widely used detectors for CuKα and softer radiations, having almost completely displaced Geiger counters for single crystal studies. Proportional counters, in turn, are being replaced by scintillation counters for many applications, but they will probably hold their own for a long time whenever their energy resolution and low non-X-ray background counting rate are important.

Proportional counter arrays

For the simultaneous measurement of several reflexions it is necessary to arrange a number of detecting elements in a line (p. 64). Three proportional counters can be stacked in the manner

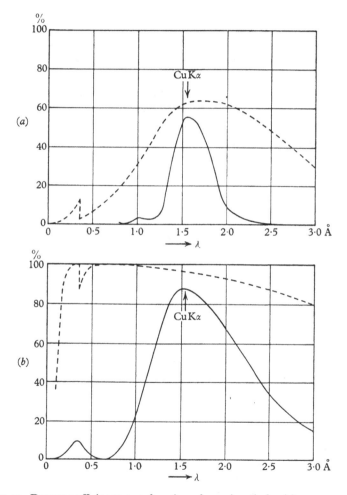

Fig. 53. Detector efficiency as a function of wavelength for (*a*) xenon proportional counter, and (*b*) scintillation counter. The dashed curves show the efficiencies of the counters themselves; the full curves show the efficiencies for CuKα radiation with a pulse-height analyser set to accept 90 per cent of the distribution curve (Parrish & Kohler, 1956).

TABLE VIII. *Performance of X-ray counters*

Detector	Useful wavelength range	Efficiency	Background counting rate	Output signal	Stability	Counting rate for counting loss < 10 %	Sensitive area
Proportional counter	CrKα to CuKα with xenon filling	With pulse height analysis (PHA) ~ 50 % for the desired X-rays, falling to ~ 25 % at an energy 7 % smaller or larger	~ 10 counts/min	~ 1 mV. High gain amplifier required	Very good	50,000 c/s with PHA; 200,000 c/s without PHA	~ 8 × 8 mm
Geiger counter	CrKα to CuKα with argon filling. MoKα with krypton filling	~ 60 %; wavelength discrimination achieved only by varying absorption	~ 60 counts/min	> 1 V	Good, especially with quenching	~ 300 c/s	~ 5 × 5 mm
Scintillation counter	CuKα to AgKα	> 85 % for all wavelengths without PHA	~ 5 counts/min without cooling	~ 0·1 V	Very good if the NaI(Tl) crystal is carefully sealed. EHT supply must have a stability of about 0·01 %	~ 5 × 10⁵ c/s	Can be very large and uniform

shown in Fig. 54 so that their windows are separated by no more than the counter radius. The absorbing paths in the outer two counters are a little shorter than in the central one and there will be a slight loss of sensitivity; the appropriate scale factors are readily determined experimentally. Phillips (1966, unpublished)

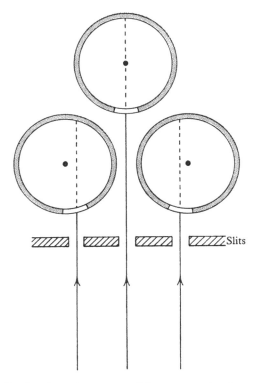

Fig. 54. Section through an array of three proportional counters for the simultaneous measurement of three reflexions.

has successfully used an array of five counters stacked in an echelon in a similar way. Alternatively, scintillation counters with light guides may be used (p. 130).

X-ray Geiger counters

When the voltage across the electrodes of a counting chamber is increased beyond the point where the gas multiplication is about

10^8 or 10^9, the discharge spreads along the whole length of the counter and its magnitude is then quite independent of the original number of ion-pairs. If a small amount of a 'quenching agent' is added to the noble-gas filling, the production of further ionization by the positive ions as they approach the outer cylinder is inhibited and the pulse lasts only until these ions are collected. Organic vapours such as ethyl alcohol were formerly employed for quenching, but halogen vapours are now generally used: halogen-quenched Geiger counters have a lower starting voltage and a more uniform sensitivity over their cross-section.

In a Geiger counter there is no need to ensure that the gas multiplication always takes place in a region of uniform field strength, as is necessary with proportional counters. Geiger counters, therefore, are usually of the end-window type and the form of construction is similar to that shown in Fig. 43 *a*. The sensitivity across the window is also sketched in the figure: the incident beam should pass near the axis of the counter without striking the central wire where the sensitivity is lower.

The filling of the counter consists of argon or krypton, at a pressure of 30–60 cm of mercury, to which about 1 per cent of halogen vapour has been added. Fig. 44 shows the absorption of different characteristic X-rays in varying columns of gas; after multiplying by the transmission factor of the window (generally about 0·7), this absorption determines the efficiency of the detector.

The stability of a Geiger counter is characterized by the length and flatness of its 'plateau', which is obtained by plotting the counting rate with a fixed incident flux against the counter voltage (Fig. 55). The form of the initial part of this curve depends upon the sensitivity of the amplifier coupled to the counter, since the 'starting voltage' is merely the point at which the pulses are large enough to be counted. Circuits coupled to a Geiger counter usually have a sensitivity of the order of 0·1 V and the pulses at mid-plateau may be a few volts in amplitude.

The sensitivity of these circuits also affects the 'dead-time' of the counter. While the positive-ion sheath from one discharge is travelling outwards, the field near the central wire is reduced below the threshold; any events occurring before the threshold field strength is again reached are not detected. The collection time of the positive ions is usually between 400 and 1,000 μs: this time interval is also known as the 'recovery time' of the counter since full-size pulses are only produced after positive-ion collection. The dead-time with typical amplifiers may be one-quarter to one-half of the recovery time; its precise value depends on the voltage applied to the counter.

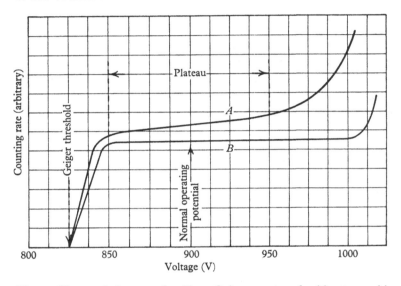

Fig. 55. Characteristic curve of an X-ray Geiger counter: A, without quenching circuit, slope 0·08 per cent per volt at mid-plateau; B, with quenching circuit, slope 0·03 per cent per volt (Arndt, 1955).

Counting loss corrections (see p. 144) can be made provided the dead-time is precisely known. For this reason it is customary to introduce an electronically defined paralysis time in the detecting circuitry which is greater than the dead-time. No counter pulses can be registered during the paralysis time following the previous count. A better arrangement consists of connecting the counter to a quenching circuit; after a discharge the voltage applied to the counter is lowered to below the Geiger threshold for a period very slightly in excess of the recovery time. The probability of further ionization as the positive ions approach the cathode is then greatly reduced and a much flatter plateau results (Fig. 55). The most effective quenching circuits are of the triggered multivibrator type (Getting, 1947; Cooke-Yarborough, Florida & Davey, 1949).

Geiger counters are still employed in X-ray powder diffractometry: in single crystal diffractometry, where the range of counting rates is much greater, they are replaced by proportional or scintillation counters. Table VIII illustrates the shortcomings of the Geiger counter in every respect save that of yielding a large output pulse.

4.3. Scintillation counters for X-rays

Developments in scintillation counters for X-ray diffractometry have been very rapid during the last few years. Since the first

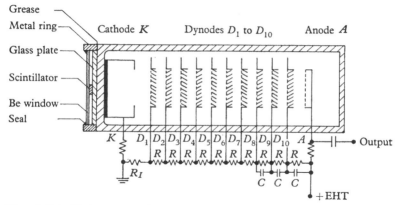

Fig. 56. Scintillation counter (schematic). In the type of construction shown here the sealed scintillator assembly is removable. The entire counter must be surrounded by a light-tight can which may also provide shielding against magnetic fields. The resistor R_I is chosen in relation to the other resistors of the chain to provide the recommended K–D_1 voltage. The decoupling condensers C are required only when the output pulses are large.

report of the use of scintillation counters for low-energy X-ray detection (West, Mayerhof & Hofstadter, 1951), these counters have been used ever more widely, and the scintillation counter, consisting of a single crystal of thallium-activated sodium iodide, abbreviated NaI(Tl), and a photomultiplier, is probably now the commonest detector used in single crystal X-ray diffractometry.

In scintillation counters, Fig. 56, the incoming X-ray quanta are absorbed in a fluorescent material, where their energy is converted into fluorescence photons with wavelengths lying in the optical region. The scintillating crystal is optically coupled to the cathode of a multi-stage photomultiplier tube where the fluorescence photons produce photoelectrons. These photoelectrons are accelerated to strike an electrode called a dynode, which is maintained at a positive potential with respect to the cathode. At this first dynode additional electrons are produced by a secondary emission process; the electron stream is further amplified at a succession of subsequent dynodes, before final collection takes place at the anode of the photomultiplier. The final gain of a typical 11-stage multiplier is between 10^6 and 10^8, depending on the dynode voltage; thus one photoelectron at the photocathode gives rise to between 10^6 and 10^8 electrons at the anode. The photomultiplier is normally connected as in Fig. 56, the chain of resistors R providing suitable accelerating voltages between the dynodes.

The number of photoelectrons generated, n, is proportional to the energy of the incident quantum: the scintillation counter, therefore, functions as a proportional device. n is given by the following relation:

$$n = (E_x/E_p) C_{xp} T_p f C_{pe}, \tag{4.4}$$

where E_x is the energy of the incident quantum and E_p that of the fluorescence photon. For NaI(Tl) the maximum of the emission spectrum is 4,200 Å and E_p is thus $\simeq$ 3 eV. C_{xp} is the fluorescence efficiency of the crystal: values of C_{xp} for common scintillators are given in Table IX. T_p is the transparency of the phosphor to its own fluorescent radiation; this is nearly unity for good crystals of the small thickness required for X-ray detection. f is the fraction of the original photons which reach the photocathode: by coating the crystal with a reflecting or, more usually, a perfectly diffusing layer on the outside faces and by optically coupling the remaining face to the front of the photomultiplier tube with a thin layer of transparent grease, this light collection factor can also be made nearly unity. C_{pe} is the photoelectric efficiency of the photocathode, that is, the ratio of the number of photoelectrons generated to the number of photons incident on the cathode. This factor is proportional to the photocathode sensitivity, and in modern photomultipliers, with sensitivities of about 40 μA/lm, the photoelectric efficiency is about 0·1. It follows then from equation (4.4) that the number of photoelectrons generated by the detection of an 8 keV CuKα quantum is about 20, and that the pulse at the anode of a photomultiplier with gain 10^6 has an amplitude of

$$\frac{20 \times 10^6 \times e}{C} \simeq 0.2 \text{ V},$$

where C, the capacity of the anode circuit, is assumed to be 20 $\mu\mu$F. These multiple-photoelectron pulses are readily distinguished from single electrons liberated at the photocathode by thermal emission.

TABLE IX. *Properties of scintillators for X-rays*

Material	C_{xp}	Maximum emission (Å)	Remarks
NaI(Tl)	0·08	4200	Very deliquescent
CsI(Tl)	0·04	4200–5700	Non-deliquescent
CaI$_2$(Eu)*	0·16	4700	Very deliquescent
ZnS(Ag)	0·28	4500	Powder only†

* New scintillator described by Hofstadter, O'Dell & Schmidt (1964).
† The amount of light collected is generally less than with the mono-crystalline scintillators.

The pulse amplitude is subject to statistical fluctuations; just as with ionization counters, the fractional spread in pulse heights corresponding to the incidence of monoenergetic X-ray quanta is proportional to $n^{-\frac{1}{2}}$, where n is now the number of photoelectrons instead of the number of initial ion-pairs. Assuming a Gaussian distribution of pulse heights, the percentage full-width-at-half-height of the distribution curve is $100(8\ln 2/n)^{\frac{1}{2}}$ or $236n^{-\frac{1}{2}}$. This width is 53 per cent when $n = 20$, which compares with a value of about 12 per cent for xenon-filled proportional counters (Fig. 57). Thus the scintillation counter, when used with a pulse-height analyser, is notably inferior to the proportional counter in wavelength discrimination, but it is still possible to use a scintillation counter with a pulse-height analyser to eliminate harmonics of the characteristic radiation. This difference in wavelength discrimination is further accentuated by the difference in the curves relating detector efficiency with wavelength (see Fig. 53).

The quantum detection efficiency, or the fraction of the incident radiation which is absorbed in the sodium iodide crystal, is very high (Table X). For the shorter wavelengths, such as AgKα and MoKα radiations, the scintillation counter is the obvious, and indeed the only practicable, detector to use. However, for CuKα and the softer radiations the proportional counter is a real alter-

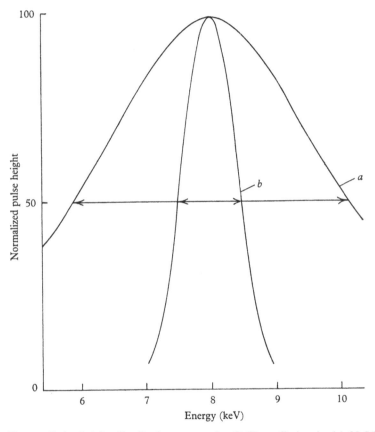

Fig. 57. Pulse-height distribution curves for CuKα radiation in (a) NaI(Tl) scintillation counter (half width 53 per cent), and (b) Xe-filled proportional counter (half width 12 per cent).

native: scintillation counters are used less, largely because of the high non-X-ray background of available counters which are not specifically designed for single crystal work.

The background counting rate in the absence of incident X-rays is plotted in Fig. 58 (curve a) as a function of the discriminator bias voltage for a typical 1 in. diameter commercial scintillation counter. This curve is an integral bias curve: N_V is the number of pulses recorded per second which are greater than V volts in amplitude. (In contrast, the pulse-height distribution curves of

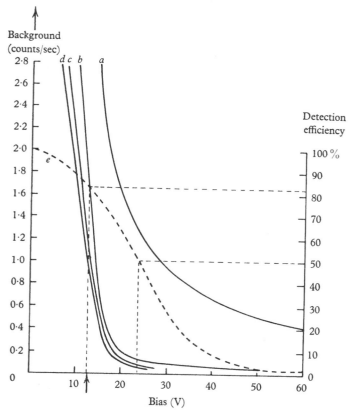

Fig. 58. Integral bias curves for scintillation counter. *a*, Background counting rate with NaI(Tl) crystal, 20 °C; *b*, background counting rate, photomultiplier only, 20 °C; *c*, background counting rate, photomultiplier only, 0 °C; *d*, background counting rate, photomultiplier only, −18 °C; *e*, normalized bias curve, CuKα X-rays. The left-hand scale refers to *a, b, c, d* and the right-hand scale to *e*.

Figs. 46 and 57 are differential bias curves, giving N_V as a function of V, where $N_V\,\mathrm{d}V$ is the number of pulses with amplitudes between V and $V+\mathrm{d}V$.) Curve *b* of Fig. 58 is the bias curve for the photomultiplier alone, with the scintillator removed. Curve *e* is the integral bias curve for CuKα, expressed as the efficiency of detection as a function of bias, and normalized to 100 per cent efficiency at zero bias. If the bias is set at the point indicated by the arrow, where the efficiency for CuKα is still 83 per cent, the background contribution of the photomultiplier alone is reduced to about 2 counts/s. This latter contribution is largely due to thermionic emission from the photocathode, and curves c and d show the effect of cooling the photomultiplier to 0 and to $-18\ ^{\circ}$C, respectively. (These temperatures can be achieved conveniently by means of compact Peltier-effect cooling devices.)

TABLE X. *X-ray absorption in NaI(Tl)*

Thickness (cm)	CuKα $\mu = 930$ cm^{-1} (%)	MoKα $\mu = 117$ cm^{-1} (%)
0·001	61	13
0·002	84	21
0·004	95	38
0·006	100	51
0·008	100	62
0·010	100	69
0·100	100	100

The difference between curves a and b is due to the general radioactive background and to cosmic radiation and is unaffected by cooling. This contribution to the background is proportional to the volume of the scintillator crystal. Curve a was obtained with a sodium iodide crystal 1 mm thick and 25 mm in diameter. However, a sensitive area 5 mm in diameter is generally adequate and a crystal 0·25 mm thick still absorbs all the incident CuKα radiation (see Table X). Thus the crystal volume, and with it the radioactive and cosmic-ray background of the counter, can be made one hundred times smaller than in the commercial detector. With a crystal of these smaller dimensions a background counting rate of 0·07 counts/s has been achieved at 0 $^{\circ}$C: the photomultiplier used was the same as that for the measurements of Fig. 58, and the discriminator bias was set for a 50 per cent detection efficiency for CuKα radiation.

The effective area of the photocathode of a photomultiplier can be reduced by providing a defocusing magnetic field between the cathode and the first dynode which steers electrons from the periphery of the cathode away from the dynode (Farkas & Varga, 1964). A short solenoid with about 250 ampère-turns, placed as indicated in Fig. 59, reduces the effective area of the cathode to about one-tenth of its normal size. The dark emission from the photocathode is reduced in the same ratio.

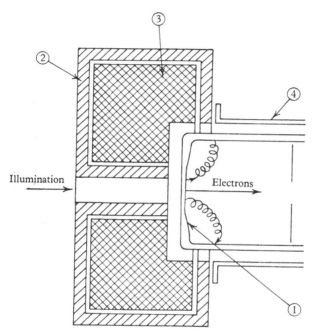

Fig. 59. Magnetic defocusing of photomultiplier: (1) photocathode, (2) iron jacket, (3) coil, (4) non-magnetic ring (after Farkas & Varga, 1964).

The light pulse produced in a scintillation counter decays with a time constant which depends on the phosphor and sometimes, for a given phosphor, on the nature of the radiation incident upon it (see p. 143). This time constant is known as the decay time of the phosphor; for NaI(Tl) it is about $2 \cdot 5 \times 10^{-7}$ s. It is customary to make the integrating and differentiating time constants of the amplifier as short as possible, or slightly longer than the decay time of the phosphor: this results in the best discrimination against noise pulses, which have a very short rise time (10^{-9} to 10^{-8} s.).

The frequency of noise pulses due to single electrons at the photo-cathode is very high—at room temperature the number per second is of the order of 10^4—and the function of the short amplifier time constants is to prevent the build-up of these noise pulses. The difference in rise time between signal and noise pulses has been used to discriminate between them (Landis & Goulding, 1964).

The properties of X-ray scintillation counters are summarized in Table VIII on p. 119.

Multiple counter arrays

Scintillation counters can be used for constructing a linear array containing a number of detecting elements for the simultaneous measurement of several reflexions (p. 63). In spite of the fact that the photomultipliers themselves are bulky it is possible to arrange the small scintillating crystals in close proximity to one another and to 'pipe' the light from scintillator to photomultiplier by means of glass or 'perspex' light guides. Fig. 60 shows an arrangement of three detectors; this can readily be extended to seven detectors by packing the conical light guides in a hexagonal array. The loss of light in the guide is about 50 per cent, and this leads to a con-siderable deterioration in the signal-to-noise ratio of the device. Some or all of the methods discussed in the previous section for the reduction of noise pulses may have to be employed to make systems of this kind usable in practice.

Continuous-dynode electron multipliers

The standard scintillation counters described earlier in this section make use of electron multiplier tubes in which the electrons are directed from one dynode to the next under the influence of an electric field. Recently, a new type of electron multiplier has been developed in which the individual dynodes are replaced by one continuous dynode. This is the channel multiplier due to Goodrich & Wiley (1962). It consists of a glass tube coated on the inside with a semi-conductor; an electric potential is applied between the ends of the tube which imparts an axial acceleration to the electrons. Further electrons are emitted from the semiconducting layer whenever an electron strikes the wall of the tube (Fig. 61). Gains in excess of 10^6 can be achieved with channels 1 mm or less in

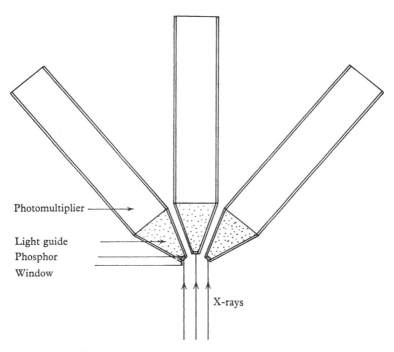

Photomultiplier

Light guide
Phosphor
Window

X-rays

Fig. 60. Array of three scintillation counters.

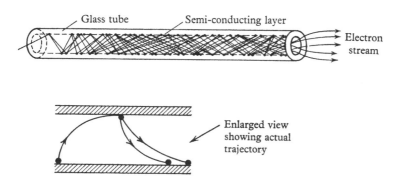

Glass tube   Semi-conducting layer

Electron
stream

Enlarged view
showing actual
trajectory

Fig. 61. The channel electron multiplier (after Goodrich & Wiley, 1962).

diameter. Channel multipliers offer the possibility of constructing very compact arrays of detectors.

Electron multipliers

X-rays can produce photoelectrons directly from suitable targets, and the electron stream can be amplified by secondary electron multiplication. Electron multipliers without photocathodes do not rely on an intermediate scintillation process and are efficient detectors for soft X-rays of wavelengths of 10 Å and more. For harder X-rays, however, they cannot compare in efficiency with scintillation counters.

4.4. Semi-conductor detectors for X-rays

Gases, by virtue of their low density, have a low linear absorption coefficient for X-rays and long columns of gas are required for reasonable counting efficiencies. It would be highly convenient to use solid detectors which could absorb almost all of the incident radiation in a very small volume. One method of using solid absorbers has been utilized for some time in scintillation counters, which have been discussed in §4.3. Here, however, a relatively large photomultiplier is required and scintillation counters are, in fact, bulkier than proportional counters.

The recent development of semi-conductor detectors in other fields has raised the attractive possibility of utilizing them for X-ray diffractometry. Compared with gas ionization counters, these detectors may be expected to have three advantages. First, they can be made extremely small and so they might permit the construction of smaller and lighter diffractometers or of diffracto-meters with multiple counter assemblies. Secondly, the mobilities of electrons and holes (the solid-state equivalent of electrons and ions in gas counters) are much greater than the mobilities of electrons and ions in a gas and the collection times are measured in tens of nanoseconds instead of tens or hundreds of microseconds: much higher counting rates can thus be achieved. Thirdly, the ionization energy of silicon is about 3·5 eV and that in other semi-conductors (for example, indium antimonide) is less than 1 eV, as compared with ionization energies in gases of about 30 eV: it might be expected that the percentage half-peak width of the pulse-height distribution curve, which is inversely proportional to the square root of the original number of ion-pairs formed, would be smaller by a factor of three or more. This last possibility, unfortunately, cannot be realized with present techniques.

Semi-conductor detectors do not yet appear to have been used in X-ray diffractometry, but advances in semi-conductor technology are so rapid that interesting developments can be confidently expected. Four types of semi-conductor detectors have so far been developed for use in other fields: intrinsic-conduction counters, diffused-junction counters, surface-barrier counters, and lithium-drifted counters. These four types are illustrated schematically in Fig. 62.

In principle, the simplest type of detector is the intrinsic-conduction counter. This consists of a very high purity ('intrinsic') silicon crystal ($\simeq 10^8 \, \Omega$ cm at 80 °K) fitted with ohmic, non-injecting electrodes. The electrons and holes created in this crystal by the incidence of an ionizing particle, or by a photoelectron produced by an incident quantum of radiation, are swept to the collecting electrodes by the applied field. This type of counter suffers from two defects, dark current and polarization. In the absence of ionizing radiation, there are still sufficient free electrons and holes to produce a dark current which would totally swamp the ionization from low-energy events; even for high-energy particles only silicon at liquid nitrogen temperatures offers any chance of being a successful detector. In addition, impurities in conduction counters produce trapping centres in which either electrons or holes, depending upon the type of impurity, are trapped; these produce a field which opposes the applied field. Even at moderate counting rates this polarization lowers the detection efficiency and destroys the proportionality between the pulse amplitude and the incident energy.

The remaining types of detector represent attempts to secure a very highly resistive material without using ultra-pure 'intrinsic' material. In diffused-junction detectors the main body of the counter consists of boron-doped p-type material containing an excess of holes. n-type material is produced near one surface by diffusing phosphorus into the bulk material. At the junction between the p- and n-type material a 'depletion region' is formed

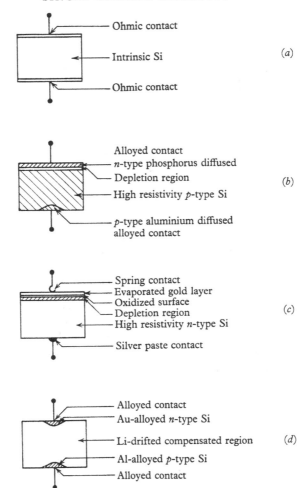

Fig. 62. Semi-conductor counters (schematic). (*a*) Intrinsic-conduction counter; (*b*) diffused-junction counter; (*c*) surface-barrier counter; (*d*) Li-drifted compensated counter.

under the influence of an electric field; in this region the free holes and electrons from the two types of material combine and produce a very high resistivity. The sensitive volume of the detector is the depletion region in which ionization is produced by the incident particle or quantum.

Surface-barrier detectors are similar in principle to diffused-junction detectors. Here the bulk material is phosphorus-doped n-type and the depletion region is formed at the junction of this and the surface onto which gold has been evaporated. This type of counter has, to date, afforded the best energy resolution and rise-time of signal.

Both diffused-junction and surface-barrier counters using silicon can be operated at room temperature. Their disadvantage lies in the limited depth of depletion region which cannot be made greater than about 1 mm, even when the purest silicon is used. A substitute for ultra-pure silicon was discovered by Pell (1960). If lithium atoms, which act as donors, are diffused into p-type silicon they remain interstitial: under the influence of an electric field they wander to acceptor sites where they are ionized and neutralize or compensate the positive acceptor ion. A sensitive region, up to 6 mm in depth, is formed in which all the acceptor sites are compensated and which behaves like ultra-pure high-resistivity intrinsic material.

For the longer X-ray wavelengths the small depletion depths of barrier and junction detectors would be no disadvantage. Table XI shows the X-ray absorption in 1 mm of silicon and of germanium. The absorption takes place primarily by the ejection of a photo-electron which is then detected in the depletion layer.

TABLE XI. *X-ray absorption in Ge and Si*

	CuKα	MoKα
Si	($\mu = 147$ cm^{-1})	($\mu = 15\cdot6$ cm^{-1})
$t = 0\cdot01$ cm	77 %	14 %
$t = 0\cdot10$ cm	100 %	79 %
Ge	($\mu = 405$ cm^{-1})	($\mu = 347$ cm^{-1})
$t = 0\cdot01$ cm	98 %	97 %
$t = 0\cdot10$ cm	100 %	100 %

The obstacle to the adoption of semi-conductor detectors for X-rays lies in the very small pulses produced. The devices function at present like ionization chambers and there is as yet nothing analogous to the Townsend avalanche which leads to gas multiplication. The amount of charge generated during the detection of an incident particle or quantum of energy E_x is

$$Q = (E_x/E_i)e,$$

where E_i is the ionization energy of the semi-conductor and $e = 1.6 \times 10^{-19}$ coulombs is the electronic charge. Thus an 8 keV CuKα quantum would produce $(8 \times 10^3)/3.6 = 2,200$ electron-hole pairs in silicon; the resulting pulse is so small as to give a very poor signal-to-noise ratio, since the various sources of amplifier noise (Goulding & Hansen, 1961) produce pulses of the same order of magnitude as the signal pulse. The largest signal is obtained when the detector capacity is as small as possible; to this end, the detector area must be kept to a minimum and it may also be necessary to have a depletion depth greater than that required on the basis of percentage absorption alone. When edge leakage effects have been eliminated by the use of a guard-ring technique (Goulding & Hansen, 1961), and all other sources of noise mini-mized by the choice of optimum amplifier time constants, the remaining noise is largely grid-current noise produced in the first valve of the amplifier; these noise pulses will be equivalent to signal pulses corresponding to an energy of about 3 keV. This ultimate limit is being approached: Bowman, Hyde, Thompson & Jared (1966) report the detection of 4.5 keV X-rays with a lithium-drifted silicon counter cooled to $-100\ °C$.

Even though semi-conductor detectors for X-rays are in their infancy, solid-state devices have been discussed here because of their certain importance in the future, both for X-rays and for neutrons. A monograph on semi-conductor detectors, such as that by Dearnaley & Northrop (1963), should be consulted for further details.

4.5. Principles of neutron detection

Many of the remarks made above on X-ray detectors apply also to neutron detectors and in the next four sections of this chapter the main emphasis will be on the differences which arise from the different properties of the two kinds of radiation.

The energy of the X-ray quanta with which we have been con-cerned above is of the order of 10 keV; it is thus sufficient to eject a photoelectron which is then detected in one of the ways described. The energy of the thermal neutrons used in most diffraction studies, on the other hand, is less than one-tenth of one electron volt. These neutrons can only be detected by their reactions with atomic nuclei; these reactions give rise to energetic particles which can then be detected by gas-ionization, electron-hole production in semi-conductors, or scintillation effects.

The nuclear reactions used in neutron detectors are shown in Table XII. The third column of this table lists the energy liberated in the reaction: this energy is very large compared with that of the thermal neutron. The total energy liberated in the detector is for all practical purposes independent of the incident neutron energy. One difference between neutron and X-ray detectors is immediately apparent: neutron counters cannot discriminate between radiation of different wavelengths by pulse-height analysis methods.

TABLE XII. *Some nuclear reactions used for the detection of thermal neutrons*

Reaction	Cross-section for 1 Å neutrons (barns)	Energy liberated (MeV)	Detected particle
^{3}He $(n, p)^3$H	3,000*	0·765	^{3}H or p
^{6}Li $(n, \alpha)^3$H	520*	0·408	α or ^{3}H
^{10}B $(n, \alpha)^7$Li	2,100*	2·78	α or ^{7}Li
^{235}U fission	320	~ 80	Fission fragments
^{239}Pu fission	410	~ 80	Fission fragments

* These cross-sections are proportional to (neutron energy)$^{-\frac{1}{2}}$ up to ~ 1 MeV.

The nuclear reaction used for the detection of the neutron results in the destruction of the neutron, and the efficiency of a given neutron detector is

$$\epsilon = p\eta, \qquad (4.5)$$

where p is the probability of the neutron reacting within the detector and η the detection efficiency for the products of the reaction. The probability p can be calculated from the cross-section of the interaction, also listed in Table XII.

Let us suppose that a beam containing n_0 neutrons per square centimetre is incident on a target having N_a absorbing centres per unit volume. At a depth x the number of neutrons is n and the number of interactions per unit area which occur between x and $x + dx$ is given by

$$-dn = n\sigma N_a dx.$$

Thus the number of neutrons emerging from a layer of thickness t is

$$n_t = n_0 e^{-\sigma N_a t}, \qquad (4.6)$$

and so $\qquad p = \dfrac{n_t - n_0}{n_0} = 1 - e^{-\sigma N_a t}$.   (4.7)

The quantity σN_a is equivalent to the linear absorption coefficient. The constant of proportionality σ has the dimensions of area and is known as the cross-section of the interaction. σ is commonly expressed in barns, where one barn $= 10^{-24}$ cm^2. In the thermal neutron range, σ, the cross-section for true absorption, is inversely proportional to the neutron velocity, and directly proportional to the neutron wavelength.

In gas ionization counters with a filling gas containing one active atom per molecule, $N_a = PL$, where P is the pressure in atmospheres and $L = 2 \cdot 7 \times 10^{19}$ cm^{-3} atm^{-1} is Loschmidt's number. For example, for a counter 10 cm long filled with ^{10}BF$_3$ at atmospheric pressure,

$$\sigma \simeq 2{,}100 \text{ barns for } 1 \text{ Å neutrons,}$$

and $\qquad \sigma N_a t = 2{,}100 \times 10^{-24} \times 2 \cdot 7 \times 10^{19} \times 10 = 0 \cdot 57;$

from equation (4.7), $p = 0 \cdot 44$. In these counters η, the detection efficiency for the energetic α-particles generated in the nuclear reaction, is unity, and hence the overall efficiency of the counter equals p.

Several types of counter make use of neutron absorbers in the form of foils, and the escaping charged particles are detected by gas- or solid-state ionization methods. If R is the range of the charged particles, only those particles which originate at a depth less than R can be counted. A solid absorber of atomic weight W, density ρ and thickness t will contain $\rho(tN/W)$ atoms per unit area, where $N = 6 \cdot 02 \times 10^{23}$ is Avogadro's number. Consequently, the number of α-particles per unit area emerging on one side of the foil will be approximately

$$n_\alpha = \tfrac{1}{2} n_0 \{ 1 - \exp - (\rho t N \sigma / W) \},   (4.8)$$

if $t < R$. It is assumed here that half the particles are ejected along the $+x$ direction into the counter. If $t > R$, R should be substituted for t in equation (4.8). Thus for a ^{10}B foil 10^{-3} cm thick ($\simeq R$), with $\rho = 1 \cdot 7$ g/c.c., $W = 10$, $\sigma = 2{,}100 \times 10^{-24}$ cm^2, we have

$$n_\alpha / n_0 = 0 \cdot 10 \quad \text{from equation (4.8).}$$

n_α/n_0 would be the approximate efficiency of a device capable of detecting all the α-particles emerging from one side of the converter foil.

Neutron diffraction experiments are often carried out in the presence of the backgrounds from γ-radiation and fast neutrons. A detector, in order to be usable for measurements of single crystal reflexions, must be sufficiently insensitive to this background to give a counting rate of at most one per second. The fast neutron flux can be reduced relatively easily by surrounding the detector with an outer shield of hydrogeneous material such as paraffin or polythene which serves to slow down fast or epithermal neutrons; a relatively thin inner shield containing boron is then sufficient to absorb most of these moderated neutrons. The γ-flux is more difficult to attenuate. The upper γ-energy limit is of the order of 10 MeV and a really efficient shield mounted on the detector arm of a diffractometer is not practicable. However, the amount of energy released in a detector when a γ-ray quantum of this energy is absorbed in the sensitive volume is generally much less than that dissipated by the charged particles produced in one of the nuclear reactions listed in Table XII, since the range of the electrons produced by the high-energy γ-ray is large compared with the counter dimensions. There is thus no problem in discriminating between thermal neutrons and γ-quanta with proportional counters: detectors such as Geiger counters which offer no energy discrimination are ruled out. The counter should, in addition, have as low an absorption for γ-rays as possible. The difference between the absorption efficiencies for neutrons and for γ-rays is large for gaseous absorbers and less for solid absorbers; consequently $^{10}BF_3$ gas-filled proportional counters are by far the most commonly employed detectors. Semi-conductor detectors and scintillation detectors are of interest mainly because of the possibility of constructing arrays, which can be used for simultaneous reflexion measurements by methods similar to those now being developed for X-rays.

The detectors for the diffracted beam should be as efficient as possible and, indeed, those described in §§4.6–4.8 can readily be constructed so as to have efficiencies in excess of 50 per cent. In contrast, the detectors used for monitoring the primary beam should have a low efficiency; they should be small, robust and stable, and capable of withstanding the high fluxes in the primary neutron beam; low-efficiency BF_3 counters or fission chambers are used for this purpose.

4.6. Gas-filled detectors for neutrons

BF_3 proportional counters

Boron-trifluoride proportional counters are filled with the gas enriched to about 96 per cent in the reactive boron isotope ^{10}B. The ionized track produced by the ^7_3Li and ^4_2He nuclei from the (n, α) reaction has a length of about 0·5 cm in BF_3 gas at 0 °C and 760 mm Hg pressure. The total energy dissipated in the reaction is about 2·3 MeV (2·8 MeV in about 7 per cent of the reactions in which no energy is lost from the counter in the form of an escaping γ-ray). Typical BF_3 counters are filled to atmospheric pressure and have a diameter of at least 1·5 cm; the whole of the kinetic energy of the charged particles is thus dissipated in the counter. If the counter is operated in the proportional region with a gas amplification A, the resulting pulse will have the amplitude

$$V = -Ane/C$$

given by equation (4.1). The number of ion-pairs produced is

$$n = (2\cdot3 \times 10^6)/33 = 70{,}000,$$

the ionization energy of BF_3 gas being about 33 eV, so that $V \simeq (A/C) \times 10^{-2}$ V, where C is expressed in picofarads. Since C, the counter capacity plus the input capacity of the amplifier is about 30 $\mu\mu$F, it is usual to operate BF_3 counters with a gas amplification of about 10 to secure a good signal-to-noise ratio. The pulses due to 1 MeV γ-rays are about 100 times smaller than the neutron pulses, so that the γ-background can very largely be eliminated.

As has been mentioned above there is no advantage to be gained by employing pulse-height analysing methods with BF_3 counters, and so it is less important than with X-ray counters to have the uniform gas amplification which results from a uniform electric field distribution. Consequently, BF_3 counters can safely be used as 'end-on' counters and advantage can be taken of the greater detection efficiency in a long column of gas. Side-window counters are used as low-efficiency monitors and in time-of-flight experiments (p. 216), in which the location of the detection event must be precisely defined.

BF_3 counters were formerly of all-metal construction, and the anode wire at the window end was supported in an internal quartz disc (Abson, Salmon & Pyrah, 1958a). The neutron beam traversed a region which contributed to the absorption without contributing to the detection efficiency. More modern counters have a ceramic end-window into which the anode is sealed directly, and the dead-volume is considerably reduced with a consequent gain in efficiency. In addition, these counters have a uniform response across the window: the older type of counter had to be carefully alined to avoid the effects of central shadowing by the insulator.

When operated at low gas amplification, sensitive BF_3 counters have a counting-life in excess of 10^{11} counts. They should not be exposed to the primary neutron beam. Eventual deterioration is probably due to the dissociation of BF_3 into BF_2 and fluorine.

^{3}He filled counters

The cross-section of the ^{3}He $(n, p)^3$H reaction is even greater than that of the ^{10}B $(n, \alpha)^7$Li reaction (Table XII). Accordingly, for the same gas pressure, ^{3}He-filled proportional counters can be made slightly smaller than $^{10}BF_3$ counters. The natural abundance of helium-3, however, is only 10^{-6} and the pure isotope is very expensive. For this reason ^{3}He counters are not widely used at present.

Boron-lined counters

The nuclear reaction which leads to the detection of the incident neutrons can also take place in solid boron. This can be introduced into a gas-filled counting chamber as a wall-coating or in the form of foils. Any of the normal counting gases, for example, helium or argon, can be used inside the chamber; the operating voltage of counters filled with these gases is generally lower than that of BF_3 counters. The efficiency of counters in which the boron is coated on the wall cannot be greater than about 13 per cent: higher efficiencies can be achieved with multiple foil counters. Thus Lowde (1950) has described a counter containing twelve parallel plates of 1 mm spacing coated with ^{10}B to a thickness of 0·5 mg/cm². The whole device measured only 58 by 30 mm. It was filled with argon at 5 atm and operated with unit gas amplification at an operating voltage of only 75 V. The efficiency of this counter was 24 per cent for thermal neutrons.

Fission chambers

Fission chambers are gas-filled counters, operated at unit gas amplification, which contain fissile material, generally in the form of U_3O_8 coated on the electrodes. Very thin coatings must be employed in order to allow the fission fragments, arising from the absorption of a thermal neutron, to escape into the sensitive volume of the detector. The mean range of these fragments in an oxide layer is about 1 μ. Fissile materials are natural α-emitters, but the pulse from a single α-particle is much smaller than that from the fission fragments which expend their energy in the detector. High concentration of fissile material must, however, be avoided in order to keep the α-activity low enough to prevent the build-up of individual α-pulses into a signal comparable in amplitude with a fission pulse.

We see then that the efficiency of fission chambers is always low. Their use in neutron diffractometry is confined to monitoring the incident beam; for this purpose the chambers are constructed in the form of parallel-plate chambers through which the primary beam passes with an attenuation of only a few per cent. Typical monitoring counters of this type have an operating voltage of about 500 V. Further details of the design and performance of fission counters have been given by Abson, Salmon & Pyrah (1958b).

4.7. Semi-conductor neutron detectors

Most semi-conductor detectors for thermal neutrons make use of the ^{10}B $(n, \alpha)^7Li$ or the ^6Li $(n, \alpha)^3H$ reactions with a target material of ^{10}B or ^6Li foil. Silicon detectors are highly transparent to thermal neutrons: it is possible to sandwich the radiator foil between two detectors, so that the incident neutron beam passes through one of the silicon crystals. With this arrangement α-particles from both sides of the foil can be detected, and the pulses from the two counters summed before amplification. In principle, it should be possible to achieve detection efficiencies of about 25 per cent with detectors of this type, but this figure has not yet been attained.

When the neutron detection takes place in a solid, as in semiconductor detectors and in scintillation counters discussed in the next section, the location in space of the detection event is much better defined than it is in long columns of gas. Thus solid-state detectors are particularly advantageous in time-of-flight experiments.

4.8. Scintillation counters for neutrons

A number of scintillators have been developed for the detection of slow neutrons (Table XIII). The most serious drawback to all these scintillators is their high sensitivity to γ-radiation. Mallett (1963, private communication) investigated a 1 in. diameter glass disc, $\frac{1}{8}$ in. thick, containing 7 per cent of ^{6}Li in contact with a photomultiplier. The whole detector was placed in an inner shield consisting of $\frac{1}{2}$ in. of boron carbide surrounded by an outer shield of three inches of polythene. At a point where the γ and fast neutron background was 5 mr/h, the background counting rate of the shielded detector was about 8 counts/s. This detector had an efficiency of greater than 80 per cent for 1 Å neutrons; it was thus inferior to a typical ^{10}BF$_3$ counter with an efficiency only slightly lower and a background counting rate under identical conditions of about 1 count/s. The background count of the scintillation counter, however, is directly proportional to the scintillator volume and in the above example this volume could have been reduced appreciably without detracting from the usefulness of the detector.

The γ-sensitivity of certain glass scintillators can be further reduced by pulse-shape discrimination methods which utilize the difference in the rise times of γ and of neutron pulses (Coceva, 1963).

TABLE XIII. *Some neutron scintillators*

Material	Literature references	Remarks
^{6}Li (Eu)	Nicholson & Snelling (1955)	—
^{10}B-plastic/ZnS(Ag)	Sun, Malmberg & Pecjak (1956); Harris (1961 a)	Poor light transparency
^{10}Borate Glass (Ce)	Bollinger, Thomas & Ginther (1962)	—
^{6}Li-Glass (Ce)	Bollinger, Thomas & Ginther (1962)	—
	Harris (1961 b); Coceva (1963)	Pulse-shape discrimination

4.9. Counting losses

All detectors have a finite resolving time: if two or more events occur within a period shorter than this time they are not resolved and only one count will be recorded and the others will be lost. The rate of arrival of X-ray quanta or neutrons at the detector is random in time. Consequently, the proportion of lost counts increases continuously with counting rate and there is no range over which the observed counting rate is strictly proportional to the incident intensity.

Corrections for counting losses are of considerable importance with X-ray Geiger counters which have resolving times of a few hundred microseconds. The other X-ray detectors and the neutron detectors which we have considered have resolving times ranging from fractions of a microsecond for some semiconductor counters to a few microseconds for large BF$_3$ counters. With the incident fluxes available from sealed-off X-ray tubes and present-day nuclear reactors the counting losses with these detectors are usually neglected. Higher flux sources are, however, coming into increasing use in both fields and the correction for counting losses is once again becoming important.

It follows from the remarks in §4.2 on pulse formation that the resolving time of a detector and its associated circuitry depends on the voltage supplied to the counter and on the sensitivity and time constants of the amplifier as well as on the discriminator or pulse-height-analyser settings. (Only for a Geiger counter is it possible to speak of a fairly well-defined 'dead time', as discussed on p. 121.)

Any discriminator or pulse-height analyser has a paralysis time which is defined by circuit component values. It is convenient to make this paralysis time slightly longer than the recovery time of the detector: the discriminator then determines the time resolution of the system and counting loss corrections can be accurately calculated.

If the observed counting rate is n_0 counts/s and the resolving time of the system is τ seconds, the system will be inactive for a period $n_0\tau$ during each second, and during this period the number of counts lost will be $nn_0\tau$, where n is the counting rate which would be observed in the absence of counting losses. We have then

$$n = n_0 + nn_0\tau,$$

or

$$n - n_0 = \frac{n_0^2\tau}{1 - n_0\tau}, \tag{4.9}$$

and the proportion of lost counts is

$$\frac{n - n_0}{n} = n_0\tau = \frac{n\tau}{1 + n\tau}.$$

It is not safe to apply these formulae when the counting losses exceed 10 per cent since, in practice, a contribution is made to the inactive period by the undetected counts also.

Form factor of the source

Many X-ray tube high-voltage supplies are only partially smoothed and the counting rate at the peak of the tube current cycle is greater than its time-average value. The counting losses are given by

$$n - n_0 = \frac{n_0^2 k\tau}{1 - n_0 k\tau}$$

(Westcott, 1948; Alexander, Kummer & Klug, 1949). k is called the form factor of the source and is defined by

$$k^{\frac{1}{2}} = \frac{\text{Root mean square intensity of source}}{\text{Mean intensity}}.$$

The form factor can be found experimentally by observing the wave-forms of X-ray tube current and voltage and making use of empirical relations between voltage and X-ray intensity (Compton

& Allison, 1935). Alternatively, k can be measured by varying τ: a series of electronically defined paralysis times must be used, each of which is greater than the resolving time of the detector. Some experimental values of k are given in Table XIV. These values of k depend on the tube voltage and waveform.

TABLE XIV

Target	Supply	k	Reference
Cu	Self-rectified tube	3·2	Arndt (1949)
		3·4	Cochran (1950)
Cu	Pulsating d.c.	1·6	Arndt (1949)
Mo	Pulsating d.c.	1·9	Cochran (1950)

Counting loss corrections when monitoring

In neutron diffractometry and sometimes in X-ray diffractometry the primary source intensity is not constant. It is then necessary to use a second counter to monitor the primary beam intensity, and the duration of each count is determined by the time taken for this monitoring counter to accumulate a given fixed number of counts. Both the measuring and the monitoring counters will be subject to counting losses. Let N_0, N_0' be the counts recorded by measuring and monitoring counters, respectively, in time t, and let N and N' be the two counts in the absence of counting losses. N_0' will be constant for a particular monitoring system. N/N' is a measure of the diffracted intensity, corrected for variations in the primary beam intensity. From equation (4.9)

$$\frac{N}{N'} = \frac{N_0}{N_0'} \frac{1 - N_0' \tau'/t}{1 - N_0 \tau/t},$$

where τ and τ' are the two resolving times.

Eastabrook & Hughes (1953) have described an arrangement which obviates counting loss corrections. They introduce electronically defined paralysis times in both measuring and monitoring channels; no count is recorded in either channel if it occurs during the paralysis time which follows a count in its own or in the other channel. The proportion of lost counts in both channels is then the same, so that the ratio of the observed counting rates is equal to that of the true counting rates.

The use of absorbing foils

If the resolving time of the system is not accurately defined by circuit values, counting loss corrections must be determined empirically using a succession of absorbing foils in the beam. In determining the absorption of the foils it is important to ensure strict monochromatism and absence of subharmonics in the beam: the proportion of harder radiation of wavelengths $\frac{1}{2}\lambda$, $\frac{1}{3}\lambda$, etc., can be greatly enhanced by the preferential absorption of the softer fundamental radiation.

Calibrated absorbing foils can also be used to attenuate the beam to a point where counting loss corrections become negligible. When the diffractometer is under on-line control by a computer, such attenuators can be inserted automatically for remeasurement under more favourable conditions, if the first measurement of a reflexion indicates an excessive peak counting rate.

4.10. Co-ordinate detectors

All the detectors which we have discussed so far are point detectors: they are used to record only beams of essentially a single wavelength which are diffracted in one particular direction. When several reflexions occur simultaneously, at best only a few of them can be recorded simultaneously, each by its own point detector.

The unit cell dimensions of the majority of crystals which are being investigated today, and the X-ray or neutron wavelengths which are being used to examine the crystals, are such that only a few reciprocal lattice points will lie on the sphere of reflexion at any one crystal orientation. This is not so with the very large biological molecules on which interest is beginning to centre. Plate I (facing p. 56) shows an extreme example. It is a *still* photograph of a crystal of poliomyelitis virus taken with CuKα radiation. Even with smaller unit cell dimensions the number of simultaneously occurring reflexions can be greatly increased by selecting a shorter wavelength: this situation is familiar in electron diffraction.

The ideal device for recording a pattern such as that shown in Plate I would be a co-ordinate detector, that is, a detector capable of recording not only every quantum incident upon a relatively large area, but also the co-ordinates within that area of the point of incidence.

Given such a detector, that volume of reciprocal space which is accessible in a normal-beam inclination setting (p. 30) could be surveyed during a rotation of the crystal about a single axis. One can visualize each diffracted quantum as having a three-number address, corresponding to the three co-ordinates of a reciprocal lattice point. In our imaginary experiment the three numbers would be the x- and y-co-ordinates of the detector and the ϕ-co-ordinate of the continuously rotating crystal at the moment the quantum is diffracted. ϕ could be recorded by attaching a digitizer to the ϕ-shaft, which would be read whenever a quantum was detected.

We note here that X-ray photographs record only two of the three co-ordinates. In a stationary film camera the ϕ-co-ordinate is lost; in a moving-film X-ray camera, such as a Weissenberg camera, ϕ is recorded, but it is necessary to keep ζ constant on a given photograph by means of a layer-line screen, so that such cameras can only record one reciprocal-lattice layer at a time.

The spatial resolution of a co-ordinate detector of the type discussed above would have to be of the order of $1,000 \times 1,000$ points. The time response could be relatively slow. The investigations in which one would like to use co-ordinate detectors are precisely those in which the intensities of individual reflexions are very small, and it should be possible to accept a maximum counting rate of between 10 and 100 counts/s for each elementary detector. The number of elementary detectors needed is such that it would be quite impracticable to provide anything akin to a normal counting chain for each element. It would not even be possible to make 10^6 electrical connexions to the array, let alone to amplify each output independently.

Two basic ways have been proposed of reducing the magnitude of the practical problem. In the first method the detectors are connected in a matrix of rows and columns and the co-ordinates of the point of detection are determined by row and column wires on which a pulse appears. In the second method the co-ordinates of the event are determined by analogue methods.

The first method is illustrated schematically in Fig. 63. Here the individual detectors, indicated by circles, are connected to a grid of wires through diodes. The row and column wires are connected to ferrite cores, amplifiers being inserted if necessary. It should be noted that only $2n$ amplifiers would be needed for n^2 detectors. The cores which are shown in the figure could be the input stages of

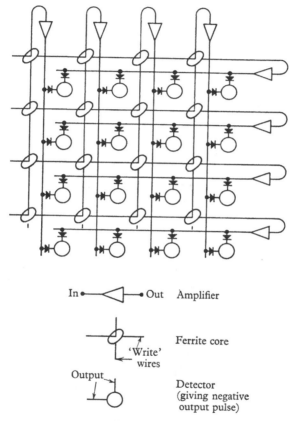

In •—◁—• Out   Amplifier

'Write' wires   Ferrite core

Output   Detector (giving negative output pulse)

Fig. 63. Matrix of detectors and ferrite cores for the recording of a large number of reflexions at a time.

individual ferrite-core counters. The capacity of the counters need only be a small number of bits since they could be read at intervals and their contents dumped into a second store containing a smaller number of longer locations: after a small number of events it would be possible to decide which elementary detectors were counting at more than background counting rate, and so corresponded to reflecting reciprocal lattice points.

One can safely suppose that advancing semi-conductor technology will make it possible to manufacture not only the detectors but also the diodes from a single silicon crystal with the help of successive photo-etch, evaporation and diffusion processes, similar to those employed today in the production of integrated solid-state circuits.

A more immediately practicable system using the same principles is being developed by Macintyre and his co-workers. Here the 'write' wires are the horizontal and vertical wires of a spark chamber which effectively consists of 18,000 elementary detectors (Cowan, Macintyre & Thomas, 1965).

In the analogue method the detector pulses are stored as charges on elementary capacitors which themselves form a two-dimensional array. The capacitors are scanned by an electron beam raster just as in a television camera. The co-ordinates of any capacitor containing a charge corresponding to the detection of a quantum are given by the timing of the read-out pulse within the raster frame period (Fig. 64).

Budal (1963) has discussed the possibilities of detector arrays of this kind, considering principally ionization counters. Arndt (1966) is pursuing a similar idea using a standard television system (Fig. 65). The diffraction pattern is formed on a fluorescent screen which is imaged, via a series of image intensifiers, on the photocathode of a television camera tube. In this camera tube a charge image of the optical image is formed on the target electrode, and this charge image in turn is scanned by an electron beam raster. If the charge corresponding to the scintillation from a single quantum can be made large enough to produce a detectable pulse as it is read out by the scanning beam, individual quanta can be counted.

Other analogue methods of determining the co-ordinates of the detected event have been proposed. Thus Bilaniuk (1960) has assembled a number of elementary semi-conductor detectors into a delay line. The point of origin of the event is given by the relative time of arrival of the pulse at the two ends of the line. Alternatively, the detector, in the form of a continuous strip semi-conductor

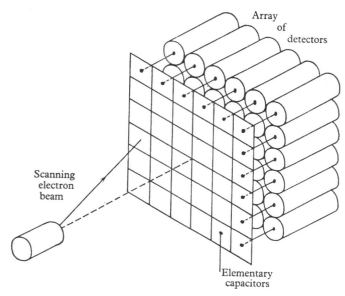

Fig. 64. Charge storage method of recording a large number of reflexions simultaneously. The detectors are connected to a two-dimensional array of capacitors which is scanned by an electron-beam raster.

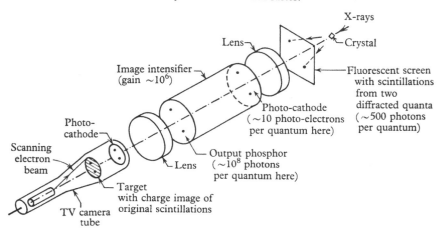

Fig. 65. The diffraction pattern is formed on the fluorescent screen. It is then amplified by means of an image intensifier and imaged on a television camera tube. With sufficient overall gain, enough charge can be deposited on the target of the camera tube from a single X-ray quantum scintillation to produce a detectable video pulse output, when the scanning electron beam discharges this target element.

counter, can be coated on its rear surface with a resistive strip which acts as a potentiometer. The position of the point of detection is given by the relative amplitudes of the pulses at the two ends of the strip. Both these methods could, at least in theory, be made two-dimensional.

Completely different methods of measuring many reflexions simultaneously are possible in neutron diffraction. Suppose that the beam incident on the crystal is polychromatic instead of

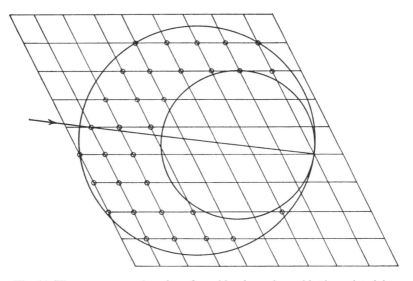

Fig. 66. The neutron wavelengths reflected by the reciprocal lattice points lying between the two Ewald spheres (shown in projection) could be determined by time-of-flight analysis. All the indicated points give rise to reflexions simultaneously.

monochromatic, as it has been in all techniques which we have described so far. This situation is depicted in Fig. 66. Here the small and the large circles represent sections through the two spheres of reflexion which correspond to the maximum and minimum wavelengths present: all those reciprocal lattice points which lie in the space between the two spheres will be in the reflecting position at any one moment. A relatively small number of discrete detectors could pick up all the diffracted beams, but it would be necessary to sort the detected quanta according to wavelengths by means of multi-channel pulse analysers. No X-ray detector can be visualized which would have the necessary energy discrimination, but with neutrons time-of-flight analysis, coupled with a pulsed source, has the necessary potentiality (p. 216).

Few of the methods mentioned have advanced beyond the drawing board. Most of them await substantial technological improvements before they become more than pipe-dreams. Yet for many problems the development of simultaneous data collection is essential and it can confidently be expected that, within the next decade, one or more of these methods will be made to work successfully.

CHAPTER 5

ELECTRONIC CIRCUITS

5.1. Introduction

A detailed discussion of the electronic circuits used in conjunction with radiation detectors would be out of place in the present monograph. The design of these circuits has become so specialized that they are no longer 'home-constructed', certainly not by the crystallographer to whom this text is primarily addressed. The aims of the present chapter are much less ambitious. They are

(1) to indicate what 'black boxes' are required in a diffractometer installation;

(2) to assist the crystallographer, who is contemplating the assembly of a complete installation from individual commercial units, in making his selection from the large range which is available;

(3) to describe the precautions necessary to secure a satisfactory performance from these units; and

(4) to indicate the amount of trouble-shooting which the user of the diffractometer may have to undertake himself.

It must be remembered that an automatic diffractometer is largely an electronic instrument and the user must have some acquaintance with electronic technology to ensure the correct functioning of his equipment. The brief notes offered in this chapter cannot supply the knowledge which can only be derived from general texts on electronics. The diffractometer user should also study more detailed discussions of detector circuitry (for example, chapter 11 of the monograph by Price, 1964). The 'logic' circuits, which are concerned with the sequence of operations of the instrument and with the setting of its shafts, cannot be properly understood without some knowledge of the principles of digital circuit design (see, for example, Pressman, 1959).

In what follows no reference will be made to the logic circuits of an automatic diffractometer. These are usually designed as a unit and the designer must try to cover as many different applications as possible. For specific applications it may be necessary to make changes in the logic for such purposes as producing the most useful measuring sequence, but there is such a great variety in the design of logic circuits that it is impossible to make useful general comments.

The detector circuits present a different situation. They can be broken down into well-defined functional units, and Fig. 67 is a block diagram which shows the individual links in a counting chain for use with the different detectors which have been discussed in Chapter 4. Individual counting chains may differ in the way in which the separate units are combined, but the overall arrangement cannot vary much from that shown here.

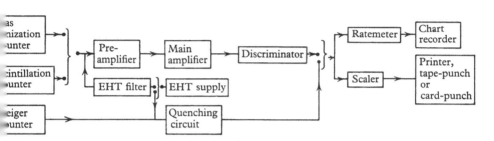

Fig. 67. Block diagram of counting chains for different kinds of detector.

5.2. Valve and transistor circuits

Nearly all the component circuits shown in Fig. 67 may make use of either valves or transistors. Transistor circuits are more compact, require less power and tend to be more reliable. Complex data-processors almost demand the use of transistor circuits. The one advantage of valves is that they lend themselves rather better to the construction of minimum-noise amplifiers. The reason for this is that gas-ionization and scintillation detectors are high output-impedance devices; thermionic valves, which are high input-impedance devices, can be coupled to these detectors rather more readily than can transistors with their relatively low input impedance. An exception to this statement is provided by the so-called 'field-effect transistor' (Radeka, 1963) which has a high input resistance. Field-effect transistors can be used as the first stage in pre-amplifiers for semi-conductor detectors and produce noise levels which are as low as, or a little lower than, those of valve pre-amplifiers (see, for example, Blalock, 1964).

The question is sometimes asked whether valve and transistor-ized circuits can be combined in one counting chain. There is no basic reason why they should not be combined; the two types of circuit will, of course, require separate power supplies. Transistor circuits, however, tend to operate with lower voltage signal levels than valve circuits: a transistorized discriminator, for example, would, in general, be designed to deal with voltage pulses of a maximum amplitude of perhaps 20 V while a corresponding valve circuit might accept pulses up to a maximum of 100 V. Since the amplitude of the pulses at the input of the amplifier is determined basically by the characteristics of the detector, it follows that amplifiers designed to work into succeeding transistor circuits can have a gain smaller by a factor of 5 than those for use with valve circuits.

Valve circuits dissipate much more power than transistor circuits. Forced air cooling is nearly always essential for a rack filled with valves; cooling is frequently unnecessary for transistors, in spite of the fact that they are more sensitive to an elevated temperature. When the two kinds of circuits are combined in one rack, the physical layout should be such that the warm air rising from the valve circuits cannot pass over the transistorized circuits.

5.3. The counting chain

Pre-amplifiers

Methods of coupling a proportional counter to the amplifier have been discussed on p. 115. It was pointed out there that the output stage of the pre-amplifier should be a cathode or emitter follower. The low output impedance of these devices prevents the attenua-tion of the signal by the shunting capacity of the cable connecting the pre-amplifier to the main amplifier.

Scintillation counters and fission chambers generate larger pulses than proportional counters, and it is frequently possible to dispense with any amplification in the pre-amplifier. The pre-amplifier may then be simply a cathode or emitter follower, with a gain just less than unity, used only for impedance matching.

The design of an appropriate pre-amplifier for semi-conductor detectors is a more complicated matter: a choice must be made between a normal 'voltage-sensitive' amplifier with a high input-resistance and a low input-capacity, and a 'charge-sensitive' amplifier (see, for example, Fairstein, 1961 a, b) with a high input-capacity. The latter type offers important advantages in some applications of semi-conductor detectors. However, for diffracto-metry it seems likely that small, low-capacity detectors will be used, and comparable signal-to-noise performance can be secured with either voltage- or charge-sensitive configurations.

When the pre-amplifier contains valves, care should be taken that the detector is affected as little as possible by heat generated in these valves.

The avoidance of 'earth-loops' in the cables connecting the pre-amplifier to the main equipment is discussed on p. 166.

Main amplifier

The overall gain required in the amplifying chain has been discussed for each detector in Chapter 4 and little need be added here to the remarks on pp. 111–14 on pulse-shaping, with special reference to high counting rates. It is worth emphasizing what the amplifier is *not* required to do. In the first instance, although quite high counting rates occur, it is not necessary to derive very precise timing pulses from the detector pulses, as must be done, for example, when coincidence or anti-coincidence techniques are employed in nuclear physics studies. Very short clipping time constants together with a wide bandwidth are not needed; indeed, an upper frequency response exceeding 1 Mc/s is not desirable since it increases the sensitivity of the amplifier to high-frequency interference. Secondly, in diffractometry there is not a strong background of higher energy radiation. Such radiation in other work can produce pulses much larger than the signal pulses and the amplifier must recover very rapidly after their occurrence; to this end, so-called non-overloading amplifiers, incorporating double shorted-delay-line pulse-shaping have been developed. The added expense of these facilities is not justifiable in diffractometry.

Discriminators, pulse-height analysers and pulse-shape analysers

With any form of detector it is necessary to count the signal pulses only and to discriminate against the smaller noise pulses. If the amplitude of the detector pulse is proportional to the energy of the incident radiation, it may be posible to discriminate also against the larger pulses, which are due to radiation of an energy greater than the desired radiation.

Accordingly, two types of pulse-height discriminators are used in diffractometry. These are lower-level discriminators which pass on only pulses greater than a pre-set minimum V, and single-channel pulse-height analysers which accept only pulses whose amplitudes are greater than a certain lower level V and smaller than $V + \delta V$. δV is the channel or window width.

It is usual for a pulse of standardized shape and amplitude to be produced for each pulse accepted by the discriminator. Thus no useful information about the shape of the input pulse can be obtained by observing the output pulse of the discriminator.

The quantities V and, in the case of a pulse-height analyser, δV can usually be varied by means of calibrated potentiometers; sometimes they can be monitored on a panel meter as well. In some analysers the two variable quantities are not V and δV, but V and $V + \delta V$, that is, the lower and the upper level of discrimination. This type of analyser is not as convenient or stable as the more usual type in which the window width is varied directly.

It will be appreciated from pulse-height distribution curves such as those of Figs. 46 and 57 that, if the whole of the distribution due to the characteristic X-radiation is to be accepted, the channel width must be at least half as great as the mean pulse height. Not all commercially available pulse-height analysers permit the channel width to be set to a width as great as this.

In order to check the performance of a detector it is frequently necessary to plot the pulse-height distribution. A narrow window width setting is needed for this purpose. The facility of sweeping the lower level automatically, in synchronism with the chart speed of a recorder connected to a counting-rate meter, is provided on some analysers, but this facility is unnecessary in a diffractometer installation. It is very useful to be able to set the channel width to 'infinity' during the setting up of a detector chain; the pulse-height analyser is then used as a simple lower-level discriminator.

Some pulse-height analysers are designed to be used at low counting rates only and are thus unsuitable in diffractometry where the resolving time should not exceed 2 μs.

As discussed on p. 137 little useful purpose is served by using pulse-height analysers with BF_3 counters, and simple discriminators are employed instead.

With scintillation counters a useful attenuation of unwanted signals can be secured by using *pulse-shape* discrimination (Owen, 1958). The shape of the output pulse from a scintillation counter depends on the decay characteristics of the scintillator for the particular radiation which is being detected. Thus for certain scintillating glasses there is a detectable difference in the shape of pulses due to neutrons and to γ-radiation (p. 143): the detection of the γ-pulses can be suppressed by pulse-shape discrimination even though the amplitudes of the neutron and γ-pulses are the same.

Another illustration of the same technique is provided by the suppression of thermal-electron noise pulses in X-ray scintillation detectors. The accidental near-coincidence of several thermal electrons can produce an output pulse whose amplitude is as great as that of pulses due to longer wavelength X-ray scintillations; the noise pulses, however, are faster than the signal pulses and Landis & Goulding (1964) have reported an almost complete suppression of noise pulses by pulse-shape discrimination.

The technique of pulse-shape discrimination has been discussed by Owen (1961). A fast timing pulse is derived from the leading edge of the output pulse of the detector; the amplitude of the output pulse is then examined at a fixed time interval after the occurrence of the timing pulse, by means of normal amplitude discriminators.

Counting-rate meters

During setting-up and alinement of the diffractometer it is essential to have an instant visual indication of the intensity incident on the detector. Counting-rate meters are employed for this purpose. Both linear and logarithmic rate meters are available in which the panel meter deflexion is proportional to the recorded intensity, or to its logarithm, respectively. It is largely a matter of personal preference which of these two types is more convenient in use.

With linear rate meters it is necessary to switch the range of the instrument frequently; the sensitivities of the different ranges should be in the ratios $1:3:10:30:100$, etc. These range switches, and the indicating meter, must be within easy reach and view of the operator when he is making adjustments, for example, to the goniometer-head arcs.

Logarithmic rate meters should cover a range of three or four decades.

The meter deflexion of a counting-rate meter fluctuates statistically when a 'constant' intensity beam is incident on the detector. If the counting rate is n counts/s and the 'time constant' of the integrating circuit of the rate meter is τ s, the standard deviation of the meter deflexion is

$$\sigma = (2n\tau)^{\frac{1}{2}}. \tag{5.1}$$

With linear rate meters it is customary to provide facilities for varying the damping of the deflexion by allowing a choice of time constants τ; the time constant of logarithmic rate meters varies automatically and continuously over the whole range of the deflexion.

It may be necessary to check the stability of the incident or diffracted beams over a prolonged period of time. For this purpose it is useful to connect a chart recorder in parallel or in series with the indicating meter.

In diffractometry counting-rate meters should only be employed for semi-quantitative work. While adequate stability is necessary, a high degree of accuracy or linearity is unimportant.

Read-out scalers

In the early days of quantum detectors the pulses from the detector (generally a Geiger counter) were counted on electro-mechanical counters, similar to or identical with the message registers used by the Post Office to count the number of telephone calls made by subscribers. If the counting rate was in excess of the few counts per second which these devices could record, the rate was 'scaled down' by electronic circuits which gave one output pulse to the counter for every 2^m input pulses, m usually being between 6 and 8. Counters today are completely electronic and almost invariably decimal, but they are still known as scalers.

Scalers used in diffractometry must be of the read-out variety, that is, the results of a count must be displayed as a printed number and often recorded also on punched paper tape or punched cards. Visual indication of the count on meters, neon lights and the like is not really necessary, but can be valuable in servicing.

An essential feature of the read-out scaler is the provision of certain lay-out symbols such as 'carriage return', 'line-feed', etc., which are inserted before and after each printed-out number. In addition, it is very useful if the output code can be changed readily. It may happen that the data produced by a diffractometer must be processed on a computer other than that originally envisaged: if the output is on paper tape the tape code must be capable of being altered, preferably without the use of a soldering iron.

The scaler should have some form of checking facility. Thus it should be possible to inject a known number of pulses, for instance from an internal oscillator, during a fixed period of time. Malfunctioning of the read-out circuits and the printer or punch can most readily be detected by setting every decade of the scaler to every digit from 0 to 9 in turn.

High-voltage power supplies

The high voltage (or *E*xtra *H*igh *T*ension, EHT for short) needed for gas-ionization and scintillation counters must be well stabilized and smoothed. The current which is drawn is usually no more than a few tens of microamperes. It should be noted, however, that the dynode-resistor chain in some commercial scintillation counters (p. 123) may have a relatively low overall resistance; these counters are then unsuitable for use with those high-voltage units which cannot supply a sufficiently large current.

Even though the high voltage is adequately smoothed it may be desirable to fit a π-section filter (Fig. 68) in close proximity to the detector in order to filter out any interference picked up in the long cables between the power supply unit and the detector.

Low-voltage power supplies

Most of the individual circuits of the counting chain are normally supplied complete with their own power supplies. It sometimes happens, however, that power must be provided for a simple circuit which is not so equipped (for example, a pre-amplifier or an inverting amplifier which must be built into an existing chassis to

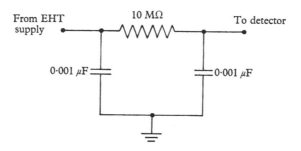

Fig. 68. π-section EHT filter.

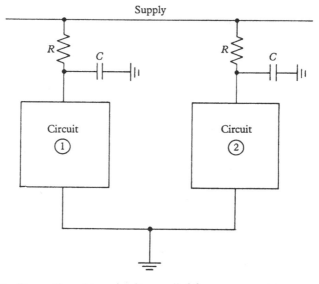

Fig. 69. Decoupling of two circuits supplied from a common power supply.

provide pulses of the correct polarity and amplitude to drive a succeeding unit). The question then arises whether a small amount of power can be drawn from one of the existing units.

Most counting circuits are fed from stabilized power supplies, that is, from supplies whose output voltage varies very little with changes in the current drawn. This is done not because these circuits are sensitive to the exact d.c. voltage of the supply, but because stabilization produces a source with a low output impedance; the coupling between different circuits or parts of the same circuit through a common source impedance is then greatly reduced. When the supply is not stabilized, and sometimes even when it is, interaction between different circuits can be considerably reduced by 'decoupling' the individual elements (Fig. 69). The decoupling resistors R should be so chosen that the voltage drop iR due to the steady current i lowers the supply voltage by, perhaps, 10 per cent. The decoupling condensers C should have a value such that the time constant $\tau = CR$ is long compared with the longest current pulse to be accepted by the circuit. (τ will be expressed in microseconds if R is expressed in ohms and C in microfarads.)

5.4. Noise and interference

The sequence of setting, measuring, printing and resetting operations in any other than a completely manual diffractometer must be controlled by sequencing circuits. These circuits frequently employ electro-mechanical devices such as relays, stepping-switches or uniselectors, and counters. All these devices contain coils or solenoids with considerable inductance; great care must be taken that the devices are adequately suppressed so that the opening of contacts in the control circuit does not produce electrical interference. If only one diffractometer is in use it is usually possible to arrange the sequence of operations in such a way that measuring operations ('counting') only commence after the most troublesome devices have operated, but if more than one diffractometer is in use this is impossible.

Inductive devices can be suppressed by connecting across them semi-conductor diodes (Fig. 70) whose polarity is such that no continuous forward current flows through the diode; only the current due to the back e.m.f. can flow through the diode, so that the electrical energy is dissipated in the ohmic resistance of a closed circuit instead of appearing as a spark across the breaking contact. The diode must have a generous voltage and current rating: the back e.m.f. in the coil may be several times greater than the supply voltage and large instantaneous currents may flow. The circulating current will, however, delay the de-energizing of the solenoid, and if rapid operation is essential the diode is best replaced by a double-ended Zener diode with a Zener voltage somewhat in excess of the supply voltage. These diodes have the property of conducting only when the voltage across them in either direction exceeds a value known as the Zener voltage; thus the normal operation of the solenoid is not affected, but the greater part of the inductive surge is suppressed. Once installed, the

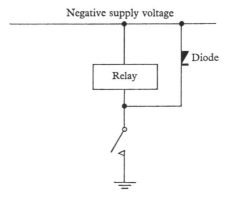

Fig. 70. Suppression of inductive surges by means of diode.

spark-suppression diodes should be checked from time to time, since the failure of a diode may become manifest only in the interference caused in other equipment.

The signal pulses from X-ray proportional counters, neutron BF_3 counters and semi-conductor detectors are usually smaller than 1 mV. One of the objects in the design of a counting chain is to produce an optimum signal-to-noise ratio and to reduce external interference to a minimum. In practice, noise and interference will be introduced only in the early stages of the counting chain where the signal level is low, that is, in the pre-amplifier and the first few stages of the main amplifier.

The theory of optimum amplifier design will not be discussed here: interested readers are referred to Gillespie's monograph (1953) on valve amplifiers, Fairstein's (1962) treatment of transistor amplifiers, and Goulding & Hansen's (1961) discussion of signal-to-noise ratio in semi-conductor detectors.

We shall suppose that a well-designed amplifier system, suitable for the chosen detector, is employed. It may happen that a system which worked well in the manufacturer's test department proves too noisy when first set up on a diffractometer in the laboratory. The first indication of trouble is an unduly high counting rate recorded on the scaler or rate-meter of the counting chain.

Such interference is generally due to one of the following causes:

(1) Too small a signal from the detector (for example, arising from too low a gas multiplication in the proportional counter), so that the overall electronic gain is so high that the inherent amplifier noise is amplified sufficiently to cause interference.

(2) Partial electrical breakdown or leakage across components in the high-tension circuit, such as condensers, plugs or sockets.

(3) High-frequency oscillations in the amplifier.

(4) External noise in the form of electromagnetic radiation from unsuppressed relays, thermostats, switches, electric motors, etc.

(5) Mains-borne interference.

(6) Mains-frequency ripple due to ineffective earthing or 'earth loops'.

A final, not altogether rare, possibility is that a strong beam of radiation is entering the detector through faulty shielding!

These various sources of interference can be recognized very easily by means of an oscilloscope attached to the output of the amplifier.

The elimination of amplifier noise (1) is obvious; troubles of the second kind can usually be localized and remedied by replacing the faulty component; oscillations in an amplifier (3) are the province of the serviceman. External noise (4) may be extremely difficult to eliminate, since the source may not be under the crystallographer's control. If, as is usual, the detector and all circuits are enclosed in earthed metal boxes and if all cables outside the rack of counting circuitry are screened, the interference is unlikely to reach the system directly; it is much more common for the one unscreened cable, which is usually the mains cable, to act as an aerial for radiated interference and to lead it into the rack or console. The mains cable and all power lines in the laboratory should, therefore, also be screened. If mains-borne interference (5) still reaches the equipment, isolating transformers, with earthed screens between primary and secondary windings, or special mains filters should be fitted. In extreme cases 'clean mains' from a separate motor alternator may be necessary, but the connexions between the generator and the equipment must still be well screened.

The most common source of trouble (6) is due to circulating currents between points which are all nominally at earth potential. There should never be more than a single 'earth' connexion between two units in a counting chain in which the signal is at a low level. This means that the detector and the pre-amplifier should be insulated from the diffractometer, the only connexion to the remaining circuits being either by a single wire or, preferably, through the screening of one of the connecting cables. All other screens should be earthed at one end only, generally at the main rack or console.

It is sometimes necessary to provide one very stout earth connexion to this rack and to take it to a copper plate buried in the ground, or at least to a water pipe, instead of to the common earth connexion of the laboratory power points.

5.5. Testing of counting chain

Test gear and fault finding

The most convenient way of locating faults is to have replacement circuits available for any units which are suspected. The modern form of modular construction means that small compact printed-circuit cards can be exchanged instead of bulky chassis. Present-day electronic circuits are very reliable and failures are more often than not in trivial components such as plugs and sockets.

An essential item of equipment is a variable-frequency variable-amplitude pulse generator. This is used to check the calibration of discriminators and pulse-height analysers as well as the calibration of rate meters and scalers.

A good oscilloscope is indispensable, and complete familiarity with its controls and method of use is extremely worthwhile. Counter and amplifier pulses are of short duration and so a high-

PLATE V

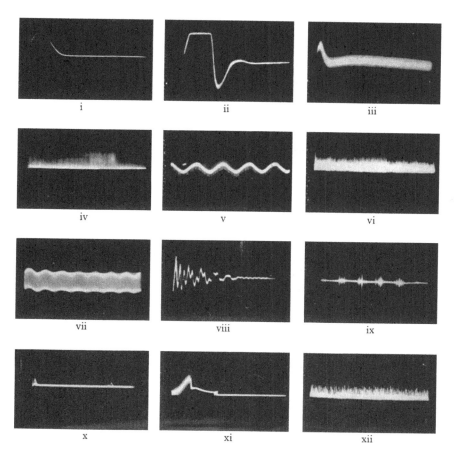

Top row. Proportional–counter pulse shapes, triggered time-base: (i) Correct pulse shape; sweep-time 6 μs. (ii) Gas-amplification too high; note saturation and overshoot; sweep-time 6 μs. (iii) Gas-amplification too low, electronic amplification too high; note poor signal-to-noise ratio; sweep-time 10 μs.

Second row. Proportional–counter pulse shapes, time-base synchronized to mains frequency; sweep-time 250 msec. (iv) Correct pulse shape. (v) Counter pulses in the presence of mains 'hum'. (vi) Counting rate too high on edge of direct beam.

Third row. (vii) Hum-modulated high-frequency oscillation; time-base synchronized to mains. (viii) High-frequency oscillation; sweep-time 6 μs. (ix) Noise from un-suppressed commutator-type electric motor; sweep-time 250 ms.

Bottom row. (x) Noise due to switching of unsuppressed relay. The sharp spike has an amplitude five times higher than the supply voltage. Triggered time base; sweep-time 100 ms. (xi) Oscillation of amplifier following a pulse ('ringing'); sweep-time 100 μs. (xii) Breakdown of EHT condenser; sweep-time 250 ms.

brilliance cathode-ray tube is needed. These pulses occur at random intervals in time and cannot be examined with a free-running time-base: the oscilloscope must have a facility for triggering the time-base from the leading edge of the pulse which is to be observed. A double-beam cathode-ray tube is not needed, since in checking detectors and detector circuitry it is rarely necessary to compare two wave-forms simultaneously.

Many crystallographers may feel that they would prefer to leave all fault finding and setting up of electronic equipment to skilled electronics technicians and to manufacturers' servicemen. Frustrating delays, however, can be avoided by some acquaintance with the circuitry; such familiarity will bring with it a greater awareness of the limitations of the equipment and a greater confidence in the experimental results.

Detector pulse shapes

Plate V shows a series of photographs of the screen of a cathode-ray oscillograph connected to the output of the main amplifier; the photographs illustrate correct and incorrect pulse shapes obtained under various conditions, as indicated in the legend. The detector was an X-ray proportional counter receiving CuKα radiation. A brief examination of displays like (i) and (iv) in Plate V can be a substitute for the lengthy plotting of a pulse-height distribution curve in checking the performance of the detector.

Setting up and periodic checks of counting circuits

When a new detector (proportional or scintillation counter) is connected to the counting chain, the manufacturer's specifications for high voltage should be strictly adhered to. In the absence of such instructions, the detector should be set to receive a weak beam of radiation and the high voltage raised slowly. The amplifier output is observed on an oscilloscope, using an amplifier gain which is sufficiently high for a considerable amount of noise to be visible: when the signal pulses appear the amplifier gain can be reduced to the desired level. After approximate operating conditions have been established, a pulse-height distribution curve is plotted: the mean pulse height should be somewhere near the centre of the range of the lower-level control of the pulse-height analyser (or of the discriminator control if a simple discriminator is used). The channel width is then set as discussed on p. 109. Finally, a check is made that small variations of the high voltage in either direction do not affect the counting rate: slightly larger variations should reduce it.

The precise operating point of all detectors changes slowly with time and the pulse-height distribution curve should be replotted at regular intervals. With X-ray proportional counters it is best to carry out the test of varying the high voltage at the beginning and perhaps also half-way through the working day, especially if the ambient temperature is subject to appreciable fluctuations. Scintillation counters are considerably more stable and tests need not be made more than once a week.

The background counting rate in the absence of an incident beam should also be checked periodically.

The count recorded by the scaler should be compared with that indicated by the rate meter, and both checked periodically with a pulse generator.

In first setting up a diffractometer, not only will the performance of the detector be unknown, but the crystal and possibly the whole instrument may be so badly alined that no primary beam reaches the detector. In this case a radioactive source can be used for checking the functioning of the detector. Suitable sources are Fe-55 (MnKα X-rays only; half-life 2·9 years) and Cd-109 (AgKα X-rays + 4 per cent of harder γ-rays; half-life 330 days) for longer-wavelength and shorter-wavelength X-ray detectors respectively, and a radium–beryllium source, surrounded by a 5–10 cm layer of paraffin, for neutron detectors. Microcuries of activity only are required.

If the source can be replaced in an exactly reproducible position relative to the detector, it can also be used to check quantitatively the performance of the detector.

CHAPTER 6

THE PRODUCTION OF THE PRIMARY
BEAM (X-RAYS)

6.1. Introduction

In this and the next chapter we shall study the methods which
are used to ensure that primary beams of the desired wavelength
and divergence fall upon the specimen. Since these methods are
quite different for X-rays and neutrons we shall here depart from
our usual procedure and consider X-rays in this chapter and
neutrons in the next.

X-ray intensity measurements are generally made with radiation
from an X-ray tube with a copper or molybdenum target. In many
investigations it is sufficient to remove the $K\beta$ component from the
direct beam by simple filtration: this procedure is discussed in
§6.4. In addition to the $K\alpha$-doublet the incident beam then still
contains a considerable proportion of non-characteristic 'white'
radiation, whose effect is suppressed by the use of balanced filters,
also treated in §6.4.

With either method of filtration the direct beam from the X-ray
target strikes the crystal. The beam divergence is determined
entirely by the dimensions of the source collimator, the crystal and
the focal spot; suitable dimensions of these are discussed in §§6.2
and 6.3.

The highest degree of monochromatization occurs when the
primary beam is reflected from a crystal monochromator. In this
case the divergence of the beam which falls on the specimen is in
part determined by the mosaic spread of the monochromator.
Monochromators are discussed in §6.5.

6.2. X-ray sources

It is not proposed to discuss the production and properties of
X-rays in the present monograph, as these subjects have been
fully covered elsewhere. A standard work on X-rays is still, after

thirty years, *X-rays in Theory and Experiment* by Compton & Allison (1935), and it is surprising how little that is totally new has been added to our knowledge of X-ray physics since the publication of this outstanding textbook. More recent work is discussed in *Physik der Röntgenstrahlen* by Blochin (1957) and in chapter 8 of the monograph by Coslett & Nixon (1960) in the present series.

The only X-ray sources which are used nowadays are thermionic-cathode high-vacuum (Coolidge) X-ray tubes. Sealed-off X-ray tubes are much the most popular type, although, as the need for fine-focus tubes and for high-intensity X-ray beams becomes greater, demountable tubes and, more particularly, rotating-anode tubes (Taylor, 1949; Mathieson, 1957; Broad, 1956) should find increasing use.

Focal spot sizes and target loadings of some typical X-ray tubes are listed in Table XV. In the next section we shall assume that the X-rays emerge from a well-defined uniformly emitting region on the target. This is a simplification. In practice, the intensity distribution within the focus is not uniform and the intensity falls off gradually at the edges; a considerable amount of radiation may originate from target regions which are outside the high-intensity focal spot itself.

TABLE XV. *Focal spot sizes and target loadings of some typical X-ray tubes*

	Focal spot (mm)	Total loading (W)	Specific loading (W/mm²)
Sealed-off tubes; stationary anode	2 × 12	2,000	83
	1 × 10	1,000	100
	0·4 × 8	800	250
Demountable tubes; stationary anode	0·1 × 6	500	830
	0·1 × 1·4	150	1,070
	0·04 dia.	20	12,000
Demountable tubes; rotating anode	0·3 × 10	6,000	2,000
	0·1 × 1	700	7,000

Nearly all X-ray tubes for crystallographic work have line foci on a flat target area which is normal to the tube axis. This line is then viewed at a small angle (1–7°) to the target face, the 'take-off angle' α, so that the fore-shortened focal spot appears approximately square (Fig. 71). The take-off angle is normally given that value which leads to a maximum intensity of the primary beam; it depends on the relative absorption of the tube-target material for electrons and for X-rays, and on the smoothness of the target face. If the whole of the resulting fore-shortened focus can be utilized the greatest intensity from a copper target is obtained at about 6°, as shown analytically and confirmed experimentally by Cole, Chambers & Wood (1962). For some experiments very small fore-shortened foci are required: they can be obtained only with a take-off angle of 1 or 2° (Guinier & Fournet, 1955; Hirsch, 1955; Furnas, 1957).

In order to obtain the greatest efficiency in producing charac-teristic radiation and the best possible ratio of characteristic to white radiation (Coslett & Nixon, 1960, §8.5; Parrish & Spielberg, 1964) the X-ray tube is supplied from a smoothed d.c. power supply rather than from a pulsating one: this has the further advantage of reducing counting-loss corrections (p. 145). Modern commercially available X-ray power supplies are usually stabilized to 0·1 per cent in high voltage and tube current, and, once the X-ray tube has reached steady operating conditions, the X-ray intensity remains constant to 0·5 per cent. The long-term stability of the X-ray source is considerably poorer; it is limited by slow changes in the absorption of the X-ray tube windows and in the self-absorption of the target. The stability of X-ray sources has been discussed by Parrish (1962).

When unstabilized X-ray sources are used, some form of inci-dent beam monitoring is necessary (§6.7).

6.3. Collimators

Fig. 71 represents the usual form of collimation for counter diffractometers, embodying a source collimator between X-ray tube focus and crystal, a backstop, and a detector collimator. We shall assume that the whole of the crystal can see the whole of the focal spot: the collimating conditions are then defined by the focal size and crystal size. The aperture S_1 limits the amount of radiation originating outside the focus itself which is 'seen' by the crystal (shown as the cross-hatched area); aperture S_2 restricts the incident beam to the cross-section of the crystal and thus prevents scattering from an unnecessarily large volume of air; S_3 is a guard aperture which just misses the direct beam but intercepts any direct X-rays originating from the edges of S_2. The backstop B intercepts the direct beam: apart from its safety aspects it has the dual purpose of further reducing the volume of air in the direct beam and of preventing this beam from entering the detector in the 'straight-through' position, with possibly deleterious effects on the detector. C_2 is the limiting detector aperture and C_1 is a guard aperture which further limits the column of air illuminated by the primary beam which

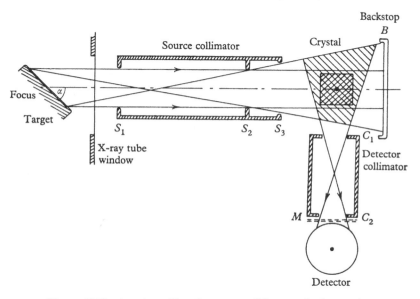

Fig. 71. Diffractometer collimating system. (Not to scale; in practice the take-off angle α is only a few degrees.)

can be seen by the counter; this column is shown shaded in Fig. 71. It is usual to provide two pairs of masks, M, by means of which the top or bottom or the right or left halves of C_2 can be covered. These masks are very useful during the alinement of the crystal.

Optimum collimating conditions

When a diffractometer is used to measure the intensities of X-ray reflexions from a single crystal, a principal aim is to adjust the geometrical conditions in such a way that the recorded intensity from the crystal is a maximum and the general background is a minimum. We must discuss, therefore, the angle through which the crystal turns near the reflecting position, the distance between focal spot and crystal, and the dimensions of the crystal and of the X-ray tube focus. It must be said at the outset that the possibilities of optimizing these quantities are much more limited with diffractometers than with X-ray cameras. The need to accommodate relatively bulky driving motors and digitizers makes it difficult to reduce the focus-to-crystal distance beyond a certain limit. In most instruments this distance is between 10 and 20 cm; it seems unlikely that it could be made much smaller. In photographic work exposure times may be reduced by scaling down the camera and its collimating system and by making use of high-brilliance microfocus tubes. Such procedures cannot be paralleled in single crystal diffractometry, but even so the question arises whether ultra-fine foci would offer any advantages when used with normal collimator lengths. Possible advantages of increasing the collimator lengths must also be considered: Cole, Chambers & Wood (1962) have shown that in certain cases this can lead to an increase in the counting rates.

The collimating conditions which lead to maximum film blackening and minimum background in photographic work have been examined by Bolduan & Bear (1949), Huxley (1953), Hirsch (1955) and Longley (1963), but their conclusions are not immediately applicable to diffractometry.

In considering collimation geometry we shall look for those conditions which lead to a maximum peak count. There is no advantage in increasing the width of a reflexion without at the same time increasing the peak count: for a constant peak count, the same total count is recorded if we integrate the area under the rocking curve in a given time, either by rocking the crystal rapidly through a large rocking range or through a smaller rocking range at a lower angular velocity. (The spot profile, or rocking curve, is the curve of counting rate versus crystal rocking angle.)

Width of the rocking curve

The angular range over which a crystal diffracts as it rotates through the Bragg position is given by

$$2\Delta = \delta_C + \delta_F + \eta + \delta\theta. \tag{6.1}$$

Here δ_C = angle subtended by the crystal at the X-ray tube focus,

$\qquad$ = $2a/s$, where $2a$ is the linear dimension of the crystal, and s the distance between focus and crystal.

δ_F = angle subtended by the X-ray tube focus at the crystal,

$\qquad$ = $2f/s$, where $2f$ is the linear dimension of the focus.

η = mosaic spread of the crystal. In defining the mosaic spread we assume that the crystal consists of mosaic blocks with a Gaussian distribution of orientation: η is the standard deviation of this distribution. For average crystals η lies between 10^{-3} and 10^{-2} radians.

$\delta\theta$ = dispersion spread,

$\qquad$ = $\tan\theta(\delta\lambda/\lambda)$, where $\delta\lambda/\lambda$ is the relative wavelength spread of the radiation used. For the $K\alpha$-doublet $\delta\lambda \simeq$ half-peak width of $K\alpha_1$ + half-peak width of $K\alpha_2 + \alpha$-doublet separation: the half-peak widths are usually small compared with the separation. For $CuK\alpha$, $\delta\lambda/\lambda = 0.00247$ and so $\delta\theta = 1.4 \times 10^{-3}$ radians at $\theta = 30°$.

It is convenient to write $\quad 2\delta = \eta + \delta\theta$

so that
$$2\Delta = \frac{2a}{s} + \frac{2f}{s} + 2\delta, \qquad (6.2)$$

where 2δ is the divergence due to the crystal and the radiation; unlike δ_C and δ_F, δ is not under the experimenter's control.

In a moving crystal method, 2Δ is the minimum angle through which the crystal must be rotated so that every mosaic block can diffract radiation covering a fixed wavelength band $\delta\lambda$ from every point on the focal spot.

Resolution of neighbouring reflexions

The rocking range affects the resolution of the diffractometer, or its ability to separate neighbouring reflexions. By differentiating Bragg's law
$$d^* = \lambda/d = 2\sin\theta$$
we get $\qquad \Delta d^* = 2\cos\theta\,\Delta\theta. \qquad (6.3)$

The maximum value of Δd^* is $2\Delta\theta$, and so reciprocal lattice points r^* r.l.u. apart will be resolved at all values of θ if

$$r^* \geqslant 2\Delta d^* \geqslant 4\Delta = 2\left(\frac{2a}{s} + \frac{2f}{s} + 2\delta\right). \qquad (6.4)$$

Conditions for maximum peak intensity

For the sake of simplicity we shall consider a circular focal spot of radius f and a stationary cylindrical crystal of radius a and length $2a$ whose cylinder axis is along the incident beam.

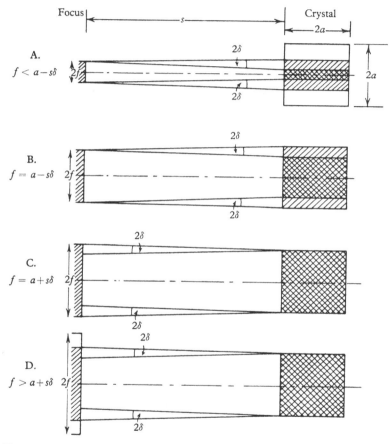

Fig. 72. Diffraction by a finite crystal from a finite focal spot. The shaded portions only of the focus are utilized by the crystal, and the shaded portions only of the crystal can diffract. The cross-hatched portions of the crystal are fully illuminated.

There are three ranges of relative focal and crystal sizes:

(1) $f < f_1$, where $f_1 = a - s\delta$ (Fig. 72A). In this range only a cylindrical portion of the crystal of radius $f + s\delta$ can diffract. When

$f = f_1$ (Fig. 72 B) the whole crystal diffracts but the outer parts of the crystal receive usable radiation from only part of the focus.

(2) $f_1 < f < f_2$, where $f_2 = a + s\delta$. In this range an increasing proportion of the focal spot is fully utilized, until, when $f = f_2$, even the edges of the crystal receive radiation over the full solid angle which can be used (Fig. 72 C).

(3) $f > f_2$. The outer edges of the focus cannot contribute to the peak count at all, and there is no point in increasing f beyond f_2.

We shall now calculate the diffracted intensity for the various ranges of f and a. Let E be the number of quanta per unit area of focus which are emitted into unit solid angle in unit time from the target; E is proportional to the specific target loading in watts per unit area.

From a region of the focal spot of radius $f < f_1$, the total number of quanta per unit time which fall on unit area of the crystal is $\pi E(f^2/s^2)$, and the total number of quanta diffracted by the crystal of cross-sectional area πa^2 is $k\pi^2 E(f^2/s^2)2a^3$, where k is a constant. The counting rate from a focal spot of radius $f_1 = a - s\delta$ is thus

$$I_1 = 2\pi^2 k \frac{f_1^2}{s^2} Ea^3$$

$$= 2\pi^2 k \frac{(a - s\delta)^2}{s^2} Ea^3. \tag{6.5}$$

Consider now an element of the focus on an annulus of radius r where $f_1 < r < f_2$. The number of usable quanta per unit time which fall on unit area of the crystal from the annulus decreases from $2\pi r\,dr(E/s^2)$ to zero as r increases from f_1 to f_2. For our purposes we can set this number equal to its mean value

$$\pi r\,dr(E/s^2),$$

so that the contribution from the outer part of the focus is

$$I_2 = \pi k 2a^3 \int_{f_1}^{f_2} \pi r\,dr\frac{E}{s^2}$$

$$= \pi^2 ka^3\frac{E}{s^2}(f_2^2 - f_1^2)$$

$$= 2\pi^2 k \frac{2a^4 s\delta}{s^2} E. \tag{6.6}$$

The counting rate at the peak of the intensity profile from a focus of radius $f_2 = a + s\delta$ is

$$I = I_1 + I_2$$
$$= 2\pi^2 k a^3 \left(\frac{a^2}{s^2} + \delta^2\right) E. \tag{6.7}$$

Let us examine the two different situations in which f and a are fixed.

(a) *No upper limit for a: focal size fixed.* When large crystals are available, and when no upper limit is dictated for the specimen size a by absorption or extinction, the specimen size should be

$$a = f + s\delta.$$

Equation (6.5) applies, so that

$$I = 2\pi^2 k a^3 \frac{f^2}{s^2} E$$
$$= 2\pi^2 k E (f + s\delta)^3 \frac{f^2}{s^2}.$$

On differentiating this expression with respect to s we find that the intensity has a minimum value when

$$s = \frac{2f}{\delta}$$

or when
$$a = 3f.$$

At very small values of s, $I \propto s^{-2}$; while at large values of s, $I \propto s$.

The explanation of the existence of a minimum is that only part of the crystal can utilize radiation from the whole of the focus. This part is cross-hatched in Fig. 72 A, and the ratio of this portion to the whole crystal volume decreases as s increases. When $a = 2s\delta$ none of the crystal can utilize the whole focal area; as s increases further, the effect of increasing crystal volume predominates. In practice s, and therefore a, cannot be made arbitrarily large, partly because of the need to keep down the physical dimensions of the diffractometer, and partly because large crystals may have excessive absorption or extinction. The lower limit on s, on the other hand, is set either by the physical dimensions of the component parts of the diffractometer, which cannot be reduced below a certain

minimum size, or by resolution considerations. We have seen
(equation 6.4), that for a desired resolution of $r*$ r.l.u.

$$\frac{a}{s}+\frac{f}{s}+\delta \leqslant \frac{r*}{4},$$

that is
$$s \geqslant \frac{8f}{r*-8\delta}, \tag{6.8}$$

when
$$a = f+s\delta.$$

When small beam divergences and a high resolution are required
very long source collimators may be advantageous; if resolution
presents no particular problems, a small value of s produces the
highest intensity. Two examples will serve to illustrate this.

Let us first consider the examination of a crystal which requires
the resolution of reflexions on an axis of 15 Å with CuKα radiation
($r* = 10^{-1}$) from an X-ray tube with a focal spot size $2f = 10^{-1}$ cm,
and let us assume a value of 2×10^{-3} for δ. The intensity minimum
will occur when $s = 2f/\delta = 50$ cm, and the minimum value of s
consistent with the required resolution is 4·5 cm. In this case it is
advisable to use the shortest practicable distance between crystal
and X-ray tube focus.

In our second example we require to resolve reflexions on a
60 Å axis ($r* = 2\cdot5 \times 10^{-2}$) with the same values of $2f = 10^{-1}$ cm
and $\delta = 2 \times 10^{-3}$ as before. The minimum value of s for the
required resolution is now 45 cm which is close to that which gives
minimum intensity. Provided sufficiently large crystals can be used
it is best to employ a very long source collimator, much longer
than 50 cm.

(b) *Fixed crystal size a.* In most cases the crystal size has an
upper limit set by the availability of suitable specimens and by the
necessity of keeping absorption and extinction small.

If a is fixed, the focal spot should be larger than the crystal, and
its optimum size will be given by

$$f = a+s\delta.$$

The appropriate equation for the peak counting rate is now (6.7)

$$I = 2\pi^2 kEa^3 \left(\frac{a^2}{s^2}+\delta^2\right).$$

The intensity, therefore, decreases with increasing s and at large values of s approaches an asymptotic value which is proportional to the crystal volume.

The value of s will be determined principally by resolution considerations. Substituting for f in (6.4) we find

$$s \geqslant \frac{8a}{r^* - 8\delta}, \qquad (6.9)$$

and when s is given its minimum value

$$f = \frac{ar^*}{r^* - 8\delta}. \qquad (6.10)$$

If we consider the same 60 Å axis crystal as above

$$(r^* = 2 \cdot 5 \times 10^{-2}; \; \delta = 2 \times 10^{-3}),$$

and suppose that the size of suitable crystals is limited to

$$2a = 4 \times 10^{-2} \text{ cm},$$

we find that the optimum focus diameter $2f$ is 1·1 mm and the minimum source collimator length is 18 cm.

Totally reflecting collimators

The refractive indices of all materials for X-rays are less than unity by an amount which is generally about 10^{-5}: thus total external reflexion, takes place when X-rays strike a polished surface at near-glancing incidence. For lead glass the critical angle for CuKα radiation is about $\frac{1}{3}°$. Glass capillary tubes have been used as collimators to increase the intensity of the beam at the specimen at the expense of increasing the divergence. (See, for example, the discussion by Hirsch (1955).) The effect of these collimators is exactly as though the focus-to-crystal distance is reduced (Jentzsch & Nähring, 1931); they may be useful when the form of construction of the diffractometer is such that this distance is unduly large.

The emerging beam is generally less uniform than one collimated by pinholes, especially if the capillary tubes are not perfectly straight or if very great care is not taken in their alinement.

6.4. The elimination of unwanted radiation

Structural studies on single crystals are almost invariably carried out using characteristic $K\alpha$ radiation. Unfortunately, in any X-ray tube the radiation characteristic of the target is always accompanied by a continuous spectrum, the 'Bremsstrahlung' or 'white radiation'. The radiation of wavelengths other than the α-doublet causes a disturbing background. Five methods exist for reducing or eliminating this unwanted radiation:

(1) The use of detectors with energy-discriminating properties. Since no known detector circuit is capable of discriminating adequately against wavelengths as close to the $K\alpha$ lines as the $K\beta$ line, this method is always supplemented by the employment of one of the following methods.

(2) The use of β-filters which reduce the intensity of the $K\beta$ line to a small fraction, usually about one per cent, of the $K\alpha$ line.

(3) The employment of balanced filters.

(4) The use of crystal-reflected radiation.

(5) Total external reflexion of the primary beam.

Pulse-height discrimination

Many of the radiation detectors used for softer X-radiation give an output pulse whose amplitude is proportional to the energy, or inversely proportional to the wavelength, of the detected X-ray quantum (Chapter 4). When such a detector is used in conjunction with a pulse-height discriminating circuit (p. 158) it is possible to reject a considerable proportion of the radiation of wavelengths other than the desired characteristic radiation. The effect of pulse-height discrimination with a xenon-filled proportional counter on the observed unfiltered and filtered spectra is shown in Figs. 73 B and D. The small peak labelled *EP* in Fig. 73 D is the escape peak (see p. 108). Figs. 73 C and and D also show the discontinuity due to the Nickel K-absorption edge. Arndt & Riley (1952) have claimed that under suitable conditions the proportion of the radiation detected with arrangement (D) which is not $K\alpha$ is about 1 per cent, but this was probably an underestimate.

For harder primary radiation ($MoK\alpha$ and $AgK\alpha$) a scintillation counter is usually employed (p. 123). The degree of pulse-height discrimination which can be obtained with these detectors is rather less than that with gas-filled proportional counters which would be too inefficient for use at shorter wavelengths. However, a considerable attenuation of the white radiation background is still possible;

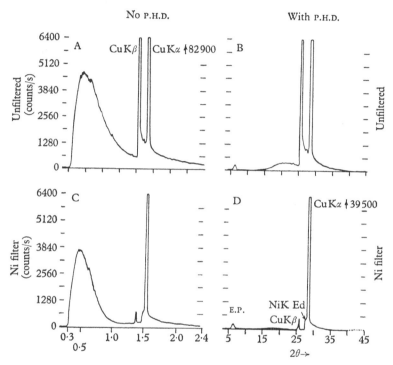

Fig. 73. Spectral composition of the radiation from a copper target X-ray tube, as detected with a Xe proportional counter: (A) unfiltered, no pulse-height discrimination; (B) unfiltered, with P.H.D.; (C) with Ni filter, no P.H.D.; (D) with Ni filter, with P.H.D. (after Parrish & Kohler, 1956).

a favouring factor is that molybdenum and silver target X-ray tubes are usually operated at voltages such that the Kα line is much nearer the peak of the continuum than with copper target tubes: the β-filter alone, therefore, removes a greater proportion of the white radiation. It should be noted that neither β-filtration nor pulse-height discrimination can do much to attenuate the white radiation of wavelength a little longer than the characteristic radiation.

As a broad generalization, it can be said that with CuKα and softer radiations accurate structure factor data can usually be collected using a β-filter and a proportional counter with pulse-height discrimination. With MoKα and harder radiations it is preferable to use balanced filters or crystal monochromators.

β-Filters

The efficacy of filters in attenuating characteristic Kβ radiation much more than Kα radiation depends on the fact that, for any X-ray target element of atomic number Z, another element, usually of atomic number Z − 1, can be found whose K-absorption edge lies between the Kα and the Kβ wavelengths of the target element. The absorption of the filter element is, much greater for the shorter Kβ wavelength than for the longer Kα wavelength.

Table XVI shows suitable β-filter materials for commonly employed target elements; the filter thicknesses required to reduce the ratio of the Kβ to the Kα intensities to 1 and 0·2 per cent are also tabulated, together with the attenuation of the Kα radiation.

Owing to the sharp discontinuity in the variation of the absorption coefficient with wavelength, filters give rise to sharp jumps in the background level.

If pure metal foils are not available, compounds of the active filter element may be used, for example Mn_2O_3 or MnO_2 instead of manganese. Methods of preparing filters have been described by many authors and a full bibliography is given by Roberts & Parrish (1962).

β-filters reduce the intensity of the white radiation of wavelengths a little shorter than that of the characteristic radiation but this reduction is limited. Fig. 73 A shows the unfiltered spectrum from a copper target X-ray tube as observed with a proportional counter; Fig. 73 C shows the effect of a Ni β-filter.

Balanced filters

The operating principle of balanced filters (Ross, 1928) is shown in Fig. 74. Two filters are constructed, generally of two elements of atomic numbers Z and Z − 1, whose absorption matches at all wavelengths other than over the narrow band between the two absorption edges λ_Z and λ_{Z-1}. If two measurements are made,

TABLE XVI. β-Filters

Target						Filter							
Element Z		$\lambda_{K\alpha_1}$ (Å)	$\lambda_{K\alpha_2}$ (Å)	$\lambda_{K\beta}$ (Å)	$\dfrac{I_{K\beta}}{I_{K\alpha_1}}$	Element		Absorption edge (Å)	μ/ρ (cm^{-1}) for target wavelengths		$\dfrac{I_{K\beta}}{I_{K\alpha_1}} = 1\%$		$\dfrac{I_{K\beta}}{I_{K\alpha_1}} = 0\cdot2\%$	
									$\lambda_{K\alpha}$	$\lambda_{K\beta}$	t (mm)	% loss Kα	t (mm)	% loss Kα
Cr	24	2·290	2·294	2·085	0·179	V	23	2·269	68·4	502	0·011	37	0·017	51
Fe	26	1·936	1·940	1·757	0·167	Mn	25	1·896	57·1	395	0·011	38	0·018	53
Co	27	1·789	1·793	1·621	0·160	Fe	26	1·743	52·8	349	0·012	39	0·019	54
Ni	28	1·658	1·662	1·500	0·187	Co	27	1·608	49·0	310	0·013	42	0·020	57
Cu	29	1·541	1·544	1·392	0·200	Ni	28	1·488	45·7	275	0·015	45	0·023	60
Mo	42	0·709	0·714	0·632	0·279	Zr	40	0·689	15·9	79	0·081	57	0·120	71
Ag	47	0·559	0·564	0·497	0·290	Pd	46	0·509	12·3	57·5	0·062	70	0·092	74

one with each filter, the difference between the two is due solely to radiation within this narrow 'pass-band'.

The situation depicted in Fig. 74 is an idealized one; in practice, even if the filters can be perfectly matched on one side of the pass-band, the balance on the other side will not be complete with only

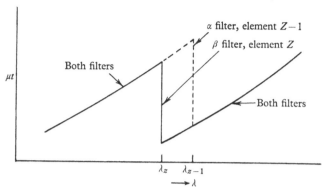

Fig. 74. The pass-band of an idealized filter pair is the wavelength range between the two absorption edges. Outside the pass-band the absorption of the two filters matches completely (after Young, 1963).

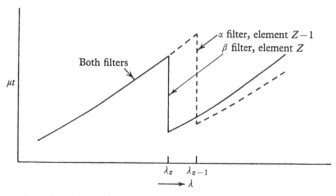

Fig. 75. A real pair in which each filter contains only one element can be balanced perfectly on only one side of the pass-band (after Young, 1963).

two filters (Fig. 75). The balance over the whole range of wavelengths can be improved by the addition of a third, neutral filter, that is, a filter which has no absorption edge near the pass-band (Kirkpatrick, 1939, 1944). The following method of balancing filters with the help of such a neutral filter is due to Young (1963).

To a first approximation the linear absorption coefficient is proportional to the cube of the X-ray wavelength; the constant of proportionality k differs for different elements and changes at an absorption edge. We can therefore write

$$\mu = k\lambda^3, \tag{6.11}$$

where μ is the linear absorption coefficient of the filter. The problem of producing a well-balanced filter-pair can be reduced to making the differences between the absorption on the two sides of the pass-band the same; a neutral filter is then made to have an absorption equal to this difference and added to the less absorbing of the pair. The balance of the composite assembly at the two edges of the pass-band is preserved at other wavelengths also, because of the similar wavelength dependence of μt for all three materials.

In the following equations, subscripts 1 and 2 refer to the two 'active' absorbers, unprimed constants to the region below the pass-band and primed constants to that above. From (6.11):

$$\mu_1 t_1 = k_1 \lambda^3 t_1, \qquad \mu_1' t_1 = k_1' \lambda^3 t_1,$$

$$\mu_2 t_2 = k_2 \lambda^3 t_2, \qquad \mu_1' t_2 = k_2' \lambda^3 t_2.$$

$\Delta(\mu t)$ and $\Delta(\mu' t)$, the differences in the absorption of the two filters below and above the pass-band, respectively, are given by

$$\Delta(\mu t) = (k_1 t_1 - k_2 t_2)\lambda^3 \quad \text{and} \quad \Delta(\mu' t) = (k_1' t_1 - k_2' t_2)\lambda^3.$$

Treating λ as constant over the pass-band, $\Delta(\mu t) = \Delta(\mu' t)$ when

$$k_1 t_1 - k_2 t_2 = k_1' t_1 - k_2' t_2,$$

in which case

$$\Delta(\mu t) = \Delta(\mu' t) = \frac{k_1' k_2 - k_1 k_2'}{k_2 - k_2'} \lambda^3 t_1. \tag{6.12}$$

We now construct a neutral filter with an absorption

$$\mu_3 t_3 = \Delta(\mu t) = \Delta(\mu' t).$$

Its wavelength dependence will be given by

$$\mu_3 t_3 = k_3 \lambda^3 t_3,$$

where k_3 is the same on both sides of the pass-band since the neutral filter has no nearby absorption edges. Thus the absorption of the neutral filter varies with wavelength on both sides of the pass-band according to the same law as the differences between the absorption of the active filters. The required thickness of the neutral filter is

$$t_3 = \frac{t_1}{k_3} \frac{k_1' k_2 - k_1 k_2'}{k_2 - k_2'}. \qquad (6.13)$$

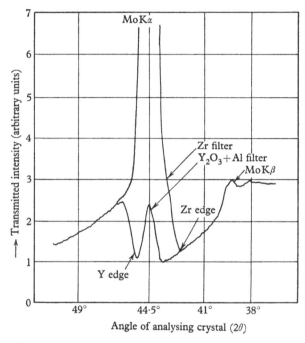

Fig. 76. Wavelength scan for a perfectly balanced Y/Zr filter pair for MoKα radiation, using an analysing crystal (after Young, 1963).

The practical process of balancing a filter pair consists of two steps:

(1) Adjusting the thicknesses of the two active filters until $\Delta(\mu t) = \Delta(\mu' t)$, with the greater transmission through the filter consisting of the element of the lower atomic number, $(Z-1)$.

(2) Adding the neutral material to this $(Z-1)$-filter and adjusting its thickness until, at any one wavelength outside the pass-band, their combined transmission is equal to that of the Z-filter.

The appropriate thickness for the β-filter (lower atomic number) is in accordance with Table XVI. Young makes the initial thicknesses of the other two filters slightly smaller than those finally required and mounts them in holders so that they can be rotated about a line parallel to the filter surface to give a fine adjustment of thickness.

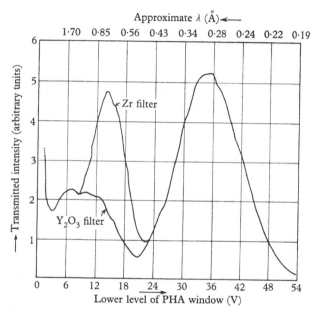

Fig. 77. Wavelength scan for the same filter pair as in Fig. 76. The wavelength of the radiation detected by a scintillation counter was varied by changing the position of the narrow constant-width window of a pulse-height analyser (PHA) (after Young, 1963).

The balance of the filter-pair must be tested at every stage by plotting the transmitted intensity as a function of wavelength for both filters. The wavelength scan can be performed by mounting an analysing crystal on the diffractometer and by varying 2θ and ω in the ratio $2:1$ (Fig. 76). A less laborious method involves the use of a detector with energy-discriminating properties in conjunction with a pulse-height analyser. The wavelength scan for balancing is achieved by varying the position of the fixed window width of the analyser (Fig. 77). Although the wavelength resolution of this type of scan is poor, it is quite adequate for the purpose.

6.5. Crystal monochromators

The highest degree of monochromatization is achieved by reflecting the direct X-ray beam from a single crystal set at the appropriate Bragg angle for the characteristic radiation. Several types of monochromators are used:

(a) *Ordinary plane monochromator crystals.* These are cut or cleaved so that strongly reflecting planes are parallel to the surface of the crystal slab.

(b) *Plane concentrating monochromators.* The surface of the crystal is cut or ground at such an angle to the reflecting planes that a concentrated beam of narrower width but greater intensity results.

(c) *Line-focusing monochromators.* These are elastically or plastically bent single crystals which may also be ground cylindrically: they are used principally in focusing powder cameras and powder diffractometers. As their name implies, they bring the X-rays diverging from the focus of the tube to a line focus.

(d) *Point-focusing monochromators.* It might be expected that these are most useful for single crystal studies, but unfortunately they are difficult to make and to adjust. Two techniques have been employed for producing a point focus. In the first method, a soft crystal, for example lithium fluoride, is deformed plastically and given a spherical or toroidal shape; in the second method, two cylindrically curved monochromators are mounted in tandem with their cylinder axes perpendicular.

(e) *Polarizing monochromators.* These produce a plane-polarized monochromatic beam which has certain specialized uses (p. 246).

Monochromator techniques have been reviewed by Roberts & Parrish (1962) who give a very full bibliography up to 1958. While crystal monochromators are used very widely in the investigation of fibres, in low-angle scattering and in powder work, they are only rarely employed in single crystal studies. Plane crystal monochromators, even concentrating monochromators, can lead to a reduction in the peak count of a Bragg reflexion by a factor of up to 10. This loss of intensity may not necessarily lead to a loss in precision in the background-corrected intensity, if the background is mainly due to white radiation (see p. 227). However, in many investigations the background is produced largely by causes other than scattered white radiation (p. 230), and there is then little advantage in using a plane crystal monochromator; these are often just those investigations where the Bragg reflexions are weak in any case and where the loss of intensity cannot be tolerated.

Focusing monochromators, especially point-focusing monochromators, are theoretically capable of producing counting rates comparable with those obtained with a filtered primary beam. However, their adjustment is frequently so critical that the intensity of the monochromatized beam drifts very readily. An account of these difficulties has been given by Teare (1960).

The intensity of an X-ray reflexion depends upon the state of polarization of the incident beam (p. 284). A crystal-reflected beam is partially plane-polarized; the resulting polarization correction depends on the perfection of the monochromator crystal and, in work of the highest accuracy, errors can be introduced as a result (p. 286).

Plane crystal monochromators

Plane monochromator crystals are about 3–5 mm on edge and about 1·5 mm thick, and must be capable of withstanding exposure to the X-ray beam and to the atmosphere. Mosaic crystals give the strongest reflexions. The reflexion used should have a low Bragg angle so as to reduce the polarization of the monochromatized beam (p. 284). If higher orders of the reflexion are weak, the harmonic contamination of the monochromatized beam is reduced; in any case, these higher orders can usually be removed by pulse-height discrimination.

Lipson, Nelson & Riley (1945) tabulate the properties of a number of crystals which have been used for monochromators. Their table has been extended by Roberts & Parrish (1962). The most generally useful crystals for plane monochromators are probably lithium fluoride, fluorite, calcite and quartz. Pentaerythritol has fallen out of favour because of its rapid deterioration in the primary X-ray beam. Crystals subject to radiation damage have been used as diffracted-beam monochromators between the specimen crystal and the detector, but when the monochromator is mounted in this position its angular acceptance range is determined by its mosaic spread and may not be sufficiently large to lead to valid intensity measurements: this situation is quite different with a normal, direct-beam monochromator in which the acceptance range is determined entirely by the angles subtended by apertures. The conditions which lead to valid intensity measurements when employing diffracted beam monochromators do not appear to have been investigated.

Monochromator crystals with a mosaic spread which is much smaller than that of the specimen crystal are very suitable for high-resolution low-angle studies, but they lead to too great a loss of intensity for normal structural studies. Monochromators employing one or more highly perfect crystals, for example germanium (Williamson & Fankuchen, 1959), fall into this category.

The beam from a plane monochromator is usually not uniform in cross-section; considerable intensity errors can result unless the specimen crystal is spherical in shape and has a very low absorption. It must also be accurately centred on the goniometer head.

Plane-crystal concentrating monochromators

When an X-ray beam is reflected from Bragg planes inclined to the surface of the crystal, the intensity of the reflected beam is enhanced by the foreshortening effect (Fig. 78). Fankuchen (1937) first proposed a plane-crystal concentrating monochromator based on this principle. This arrangement was further discussed by Bozorth & Haworth (1938) and by Fankuchen (1938). Evans, Hirsch & Kellar (1948) showed that, because of absorption effects, the gain of these monochromators, as compared with non-concentrating monochromators, is always less than two.

Line-focusing monochromators

Curved crystal monochromators which bring the X-rays diverging from the focal spot of the X-ray tube to a line focus are widely used with focusing powder diffractometers (Fig. 79). Intense Debye–Scherrer lines are produced because of the relatively large angular aperture of the beam diverging from the tube target (1–3°). Some geometrical aberration in the focus is produced by monochromators of the Johann type (Johann, 1931) in which the monochromator crystal is bent into a cylindrical form without further shaping; these aberrations are reduced by bending the crystal to a radius $2R$ and then grinding the crystal surface to a radius R (Fig. 80) (Johansson, 1932). With single crystal specimens there is no gain in recorded intensity if the divergence of the beam

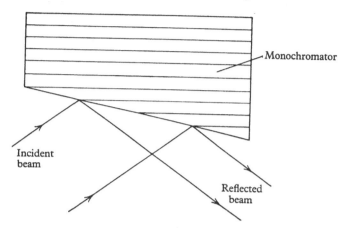

Fig. 78. The principle of the Fankuchen concentrating monochromator (after Evans, Hirsch & Kellar, 1948).

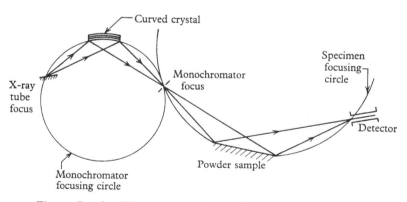

Fig. 79. Powder diffractometer with curved crystal monochromator.

incident on the specimen increases beyond 2δ, the sum of the mosaic spread and the dispersion spread of the radiation. 2δ is normally much less than 1°, and so the principal advantage of line-focusing monochromators, that of being able to utilize a beam of large angular aperture, is lost in single crystal work.

Point-focusing monochromators

Numerous arrangements have been devised for focusing a monochromatic beam to a point (see references cited by Roberts &
Parrish, 1962; Atkinson, 1958; Sandor, 1964; Elliott, 1965). In
theory, many of these should be capable of producing monochromatic beams of X-rays suitable for single crystal diffractometry.

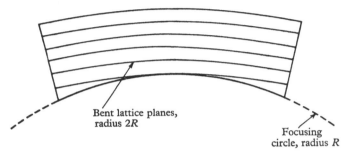

Fig. 80. Johansson line-focusing monochromator. The source, the monochromator focus and the monochromator surface lie on the focusing circle.

In practice, the performances quoted by many authors have been
disappointing: this is due to the difficulties of manufacturing,
alining, and maintaining in alinement, monochromators of this
type. Point-focusing monochromators have been used successfully
in the study of diffuse reflexions from single crystals, but have not
yet been employed with diffractometers for structural studies.

Polarizing monochromators

Monochromators are generally operated at small Bragg angles in
order to reduce the polarization of the monochromatized beam:
there is a severe intensity loss if a largely plane-polarized X-ray
beam is diffracted from a specimen crystal at Bragg angles near
45°. There are cases, however, when a plane-polarized beam is
needed, notably when an experimental correction for extinction is
made using Chandrasekhar's method (1956, 1960a, b; see p. 246).
In the experimental arrangement described by Chandrasekhar &
Phillips (1961), the 311 reflexion from diamond was used with
CuKα radiation for which $2\theta_M = 91° 34'$. The reflected X-rays
were almost completely (p. 284) polarized with the electric vector

perpendicular to the plane of incidence. The X-ray tube and the diamond monochromator ($8 \times 7 \times 2\cdot5$ mm with (111) faces) were mounted on a disc which could be rotated about a horizontal axis collinear with the collimator of the diffractometer (Fig. 81). The plane of polarization of the beam incident upon the specimen crystal could thus be rotated about the beam direction.

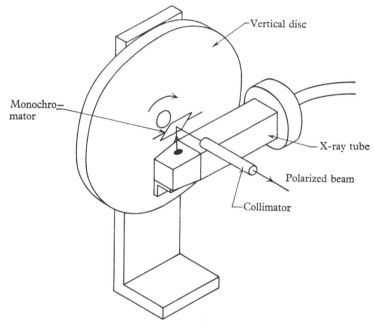

Fig. 81. Chandrasekhar's polarizing monochromator. The X-ray tube is attached to a vertical disc which can rotate about a horizontal axis. The primary beam is reflected from a diamond monochromator ($2\theta_M = 91\cdot5°$) and the reflected beam is along the horizontal axis of the disc.

Chandrasekaran (1956) adopted a similar method to obtain a polarized beam, using the 311 reflexion from a single crystal of copper ($2\theta_M = 90°\ 12'$ for CuKα radiation).

A more convenient device for producing a polarized X-ray beam has been described by Cole, Chambers & Wood (1961). Use was made of the anomalous transmission effect discovered by Borrmann (1941, 1950; see also v. Laue, 1960, p. 403). According to the dynamical theory of diffraction by a perfect crystal, there is a net energy flow along the atomic planes and two standing wave patterns perpendicular to the flow. In a simple structure, such as the diamond structure of germanium, the nodes of one standing wave pattern coincide with the atomic sites and photoelectric absorption cannot take place since the electric field strength is zero at the only points where there is any absorbing matter. Even for the anomalously transmitted beam the state of polarization with the electric vector in the plane of incidence is absorbed more readily than the polarization state with the electric vector perpendicular to the plane of incidence. By choosing the thickness of the crystal

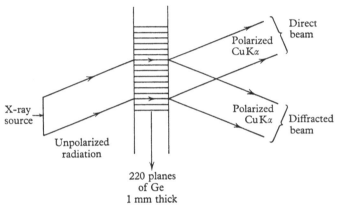

Fig. 82. Basic diffraction geometry for anomalous transmission of X-rays (Borrmann effect). The 220 planes in a germanium crystal are used to produce the polarized and monochromatized transmitted beams. Either direct or diffracted beam can be used (after Cole, Chambers & Wood, 1961).

appropriately one polarization state is suppressed and the direct and diffracted beams are polarized (Fig. 82). The effect is destroyed if the crystal is not perfect.

Cole, Chambers & Wood used a dislocation-free germanium crystal about 1 mm thick, cut to diffract CuKα radiation from the (220) planes. By utilizing the anomalously transmitted direct beam, the plane of polarization could be rotated by rotating the crystal, while the beam itself stayed fixed in space. The chief disadvantage of this type of monochromator is the very feeble beam which it produces, arising from the need to use a perfect crystal with an angular reflecting range of only a few seconds of arc.

6.6. Total external reflexion

The refractive index for X-rays is given by

$$n = 1 - x, \tag{6.14}$$

where x, which is of the order of 10^{-5}, is proportional to the square of the wavelength. Total external reflexion, therefore, takes place at sufficiently small glancing angles. The critical glancing angle θ_c for total reflexion ranges from a few minutes to a few tens of minutes of arc.

θ_c is given by

$$\theta_c = (2x)^{\frac{1}{2}} \tag{6.15}$$

and so

$$\theta_c \propto \lambda. \tag{6.16}$$

Consequently, according to the simple theory, if an optically flat surface is set at the critical glancing angle for radiation of wavelength λ_c, then only radiation for which $\lambda < \lambda_c$ will be reflected, and radiation for which $\lambda > \lambda_c$ is absorbed in the reflector.

Unfortunately, owing to absorption effects, for real reflectors the critical angle is not a perfectly defined quantity, and the ratio of the reflected to the incident intensity falls from nearly 100 per cent to zero over a finite angular range: accordingly, there is no sharp wavelength cut-off of the reflected radiation. Nevertheless, a considerable attenuation of unwanted radiation can be achieved by optical reflexion of X-rays. This subject has been studied by Ehrenberg and his co-workers (Ehrenberg, 1949; Ehrenberg & Franks, 1952; Franks, 1955), and by Henke & DuMond (1953, 1955). A point-focusing system employing gold-coated toroidal mirrors has been described by Elliott (1965). The early work on total external reflexion was reviewed by Compton & Allison (1935) who gave numerous references.

6.7. Primary beam monitoring

We have discussed those factors which affect the variation of the incident X-ray beam intensity in space; a few words must be included concerning intensity variations with time.

Modern commercial X-ray generators embodying sealed-off X-ray tubes are normally stabilized so that the X-ray intensity at the X-ray tube window varies by much less than 1 per cent over periods long enough to collect intensity data from any one crystal. On the other hand, X-ray generators making use of continuously pumped special-purpose tubes, such as fine-focus or high-brilliance rotating anode tubes, are less well stabilized: moreover, the vacuum inside such tubes is frequently inferior, and X-ray intensity fluctuations due to evaporation of the tube filament and contamination of the target, as well as those due to changes in the vacuum, are more severe.

When continuously pumped tubes are employed some form of monitoring of the primary X-ray beam may be desirable. Diffracted intensity measurements are then expressed, not as the intensity recorded in a fixed time, but as the ratio of the diffracted to the incident energy. The X-ray intensity can be monitored either on the same tube-window from which the incident beam is taken or on another window. The latter method requires no modification of the normal collimating system of the diffractometer but it is considerably less reliable; if the views of the X-ray tube focus as seen by the measuring and the monitoring system are not identical, intensity fluctuations due to focal spot movements are not adequately corrected.

Several systems have been used in which the collimated X-ray beam is monitored before it strikes the specimen. Hargreaves, Prince & Wooster (1952) passed the primary beam through an air ionization chamber placed in the collimating system. Arndt, Coates & Riley (1953) placed a cold-worked aluminium foil in the primary beam: the monitoring counter recorded a broadened powder diffraction line from this foil. Holmes (1964, private communication) inserted a small piece of tape-recorder magnetic tape in the incident, monochromatized CuKα beam and detected the iron-fluorescent radiation from the iron oxide coating. A number of monitoring arrangements, and the precautions necessary to secure adequate stability, have been discussed by Gillam & Cole (1953) and by Williamson & Smallman (1953).

A method closely akin to actual monitoring of the direct beam consists of deriving an electrical signal from the monitoring channel which is then used to stabilize the X-ray generator, generally via the filament supply circuit.

In any monitoring method it is necessary to ensure that the number of counts recorded by the monitoring counter during a measuring period is large enough to yield a small fractional standard deviation in the monitor count (p. 274). Fatigue effects in the monitor detector at high counting rates are a possible source of inaccuracy. For this and other reasons, monitoring systems are liable to produce more trouble than might be anticipated. In any experimental investigation it is always worthwhile to examine the errors in intensity measurements due to causes other than variations in incident beam intensity (Chapter 9) before deciding to embark on some form of monitoring.

CHAPTER 7

THE PRODUCTION OF THE PRIMARY
BEAM (NEUTRONS)

Quantitative neutron diffraction studies were not possible until 1945 with the advent of the nuclear reactor as a powerful source of neutrons. A high-flux reactor, such as the Harwell *Dido* or *Pluto* research reactors, has a central flux of about 10^{14} slow neutrons/cm²/sec. These neutrons move in all directions and only a proportion of about 1 in 10^5 travel in the right direction down the collimator; of these collimated neutrons, in turn, a fraction of between 10^{-3} and 10^{-2} has the right wavelength to be reflected by the monochromator. The collimated flux of monochromatic neutrons striking the sample is, therefore, 10^6 to 10^7/cm²/sec. This compares with a flux exceeding 10^{10} quanta/cm²/sec at the sample in the X-ray case. To compensate for this disparity in the incident flux and for the smaller cross-section for scattering of neutrons as compared with X-rays (Bacon, 1962), larger samples are used in neutron diffraction and the time required to count the diffracted neutrons is usually made longer.

In this chapter we shall describe briefly the collimators and monochromators used for the production of the primary neutron beam striking the sample. Only those points are discussed which relate to single crystal diffractometry: other neutron techniques are covered in Chapter 4 of G. E. Bacon's *Neutron Diffraction* (1962) and in Chapter 3 of *Thermal Neutron Scattering* (1965), edited by P. A. Egelstaff. In §7.3 on monochromators we include a discussion of the resolution of neutron reflexions, as this is closely connected with the properties of monochromators.

7.1. Neutron collimators

The size and complexity of neutron collimators are in marked contrast with the X-ray collimators described in the previous chapter. Neutron collimators are large because they must extend from the inner part of the reactor to the outside of its biological shield. Their complexity arises from a number of causes. In high-flux reactors the inner section of the collimator must be cooled: in the absence of cooling this section would be raised to a high temperature by the intense γ-radiation near the reactor core. There is a need to shut off the direct beam (for instance, by means of a floodable water switch), and to seal the collimator inside the reactor to prevent the ingress of atmospheric argon, which would be made radioactive by the high flux of neutrons.

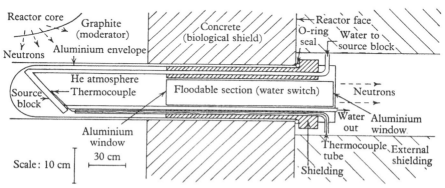

Fig. 83. Sketch of neutron collimator used with *Pluto* reactor. Access to the reactor face is possible only when the reactor is shut down and the water switch is flooded.

Fig. 83 is a diagram of a collimator designed for single crystal diffraction work at the Harwell *Pluto* reactor. The collimator does not point directly at the reactor core, and so a 'source block', consisting of a water-filled aluminium box, is installed at the inner end of the collimator; the water scatters slow neutrons in all directions and is the effective neutron source. In many reactors the collimators point at the core itself which is then the effective source.

Let us assume that the distance from the source to the outside of the biological shield is 4 m and that the cross-section of the beam as it passes down the collimator is 5×5 cm. The angular divergence of the beam, in both the horizontal and vertical planes, is 1.25×10^{-2} rad, and so the neutrons emerge from the collimator at angles of up to $0.7°$ from the axis of the collimator. The fraction of neutrons passing down the collimator is

$$\frac{1}{4\pi}(1.25 \times 10^{-2})^2,$$

or about 10^{-5}.

The most appropriate value for the collimation angle depends on the nature of the crystal under investigation. The figure of $\pm\frac{1}{2}°$ represents a reasonable compromise between the opposing requirements of high intensity and good resolution of the reflexions from crystals with unit cell dimensions less than 10 Å, examined with a neutron wavelength of about 1 Å. For crystals with larger unit

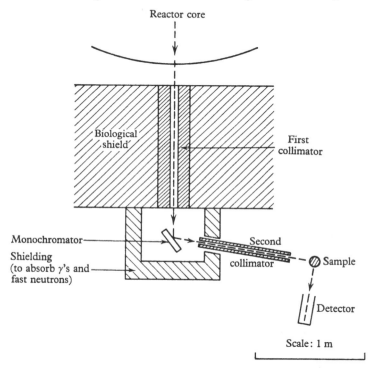

Fig. 84. Schematic diagram of neutron diffraction assembly. A second collimator is used after the monochromator, in case the degree of primary collimation is insufficient.

cells, tighter collimation or a longer wavelength are necessary to avoid overlapping of reflexions from adjacent levels of the reciprocal lattice. It is an advantage, therefore, to provide variable collimation of the beam striking the sample. One method, adopted in the Brookhaven High Flux Beam Reactor (Kevey, 1964), is to have a rotary collimator arrangement, whereby any one of three collimators of different cross-sections can be selected by remote control, without having to remove the collimator assembly as a whole from the reactor. An alternative procedure is to use an in-pile collimator for primary collimation of the beam, and to insert a second collimator *after* the monochromator (Fig. 84) to reduce the beam divergence further. The second collimator consists simply of an open channel, which is shielded from the main gamma and fast-neutron radiation, and so can be interchanged readily with conical channels of different dimensions, in order to alter the divergence of the incident beam.

7.2. Neutron spectrum emerging from collimator

In the centre of the reactor the slow neutrons are in thermal equilibrium with atoms of the moderating material, and the velocity distribution of the neutrons follows the Maxwellian law

$$n(v) = \frac{4N}{\sqrt{\pi}} \frac{v^2}{v_0^3} e^{-v^2/v_0^2}. \tag{7.1}$$

Here $n(v) \, dv$ is the number of neutrons per unit volume with a speed lying within the range v to $v + dv$, and N is the total number of neutrons per unit volume. v_0 is the most probable velocity, corresponding to the maximum value of $n(v)$ in (7.1), and is given by

$$v_0 = \left(\frac{2kT}{m}\right)^{\frac{1}{2}},$$

where k is Boltzmann's constant, T the moderator temperature and m the neutron mass.

The neutron flux $\phi(v)$ emerging from unit area of the collimator is v times the neutron density, that is,

$$\phi(v) = v n(v) = \frac{4N}{\sqrt{\pi}} \frac{v^3}{v_0^3} e^{-v^2/v_0^2}. \tag{7.2}$$

To express the neutron flux in terms of the wavelength λ, we use the de Broglie relationship
$$\lambda = h/mv, \tag{7.3}$$

where h is Planck's constant, together with the expression

$$\phi(v) \, dv = -\phi(\lambda) \, d\lambda, \tag{7.4}$$

where $\phi(\lambda) \, d\lambda$ is the neutron flux crossing unit area with a wavelength lying within the range λ to $\lambda - d\lambda$. Substituting (7.3) and (7.4) in equation (7.2) gives

$$\phi(\lambda) = \frac{4N}{\sqrt{\pi}} \frac{h}{m} \frac{\lambda_0^3}{\lambda^5} e^{-\lambda_0^2/\lambda^2},\qquad(7.5)$$

where

$$\lambda_0 = \frac{h}{mv_0} = \frac{h}{(2mkT)^{\frac{1}{2}}}.$$

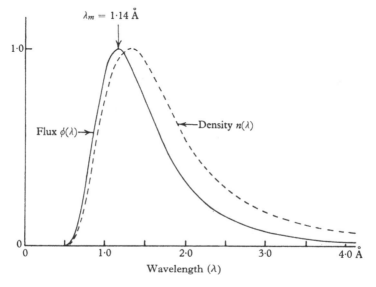

Fig. 85. Wavelength distribution of neutrons in equilibrium with a moderator at temperature $T = 20\ °C$. Full curve is neutron flux emerging from collimator and broken curve is neutron density in reactor core: both curves are normalized to a peak value of unity. The curves are shifted towards shorter wavelengths for higher values of the moderator temperature.

The maximum value of $\phi(\lambda)$, found by differentiating (7.5), occurs at a wavelength λ_m, given by

$$\lambda_m = \frac{h}{(5mkT^{\frac{1}{2}})}.$$

Fig. 85 shows the calculated neutron flux $\phi(\lambda)$ for a moderator temperature of 20 °C. In neutron diffractometry we are concerned mainly with the distribution $\phi(\lambda)$ and with the wavelength λ_m. A typical moderator temperature is in the range 20–100 °C: at 20 °C, λ_m is 1·14 Å and at 100 °C, λ_m is close to 1·00 Å.

7.3. Crystal monochromators

The selection of a narrow wavelength band from the continuous distribution shown in Fig. 85 is readily accomplished with a crystal monochromator. Crystals are chosen for their high intrinsic reflectivity Q (p. 277) and large mosaic spread. If the monochromator is a perfect crystal (for example, germanium), it selects neutrons of a fixed wavelength from a very narrow angular range of the incident beam. On the other hand, for a mosaic crystal, neutrons are selected over an angular range corresponding to its mosaic spread: the total reflected intensity is several orders of magnitude higher than for a perfect crystal (Fig. 86). Ideally, the

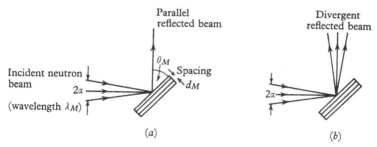

Fig. 86. Increase of reflectivity by using a mosaic crystal (b) instead of a perfect crystal (a) as monochromator. In (b) the full angular range $\pm \alpha$ of the incident beam is reflected at the wavelength $2d_M \sin \theta_M$, provided the mosaic spread η_M exceeds α. A parallel beam is reflected by a perfect crystal, (a). (The diagram applies only to the single wavelength $\lambda_M = 2d_M \sin \theta_M$; other wavelengths are reflected as shown in Fig. 88.)

mosaic spread of the monochromator should match the angular divergence of the incident beam, so as to give maximum intensity at a resolution which is still governed primarily by the collimator divergence.

Attempts have been made to produce crystals with a controlled mosaic spread, but only limited success has been attained. Thus Shull (1960) has described experiments on Si and Ge, in which nearly perfect single crystals were distorted to improve their reflectivities. The reflecting range was increased by curving the crystals at high temperature in the plastic state, followed by a straightening treatment. Appreciable increases in reflectivity were achieved, but the distortion was not uniform and so the beam intensity was not constant across the reflecting area of the crystal. Other experiments were reported by Barrett, Mueller & Heaton (1963), who introduced imperfections in several ways into neutron transmission monochromators of germanium. The recommended technique was uniaxial compression along the [110] direction at 650 °C to reduce the thickness by 2·5 per cent; this treatment improved the efficiency of the monochromator by a factor of 25–40 over that of an undeformed slab. Table XVII, taken from the paper of Barrett, Mueller & Heaton, compares the performance of a hot-pressed Ge monochromator used in transmission with a Cu monochromator used in reflexion, both set for a wavelength of 0·98 Å. The monitor rate in the table is the neutron flux recorded by a low-efficiency monitoring counter in the beam reflected by the monochromator, and the peak intensity and integrated intensity are those recorded for the 002 reflexion of a specimen crystal of alpha-uranium. The monitor rate and peak intensity are somewhat higher for the Cu monochromator, even though the integrated intensity is less; it is the peak intensity which is important, so that the comparison favours the copper monochromator rather than the germanium.

TABLE XVII. *Comparison of a Ge monochromator used in transmission and Cu monochromator used in reflexion (wavelength = 0·98 Å)*

Mono-chromator	Crystal condition	Relative monitor rates	Relative peak intensities	Relative integrated intensities
Ge	Hot pressed	63	56	174
Cu	As grown	100	100	100

Table XVIII lists a number of single crystals used as neutron monochromators, with comments indicating any special features.

In correcting for the effects of extinction or simultaneous reflexions (see Chapter 9), it is often useful to repeat measurements at different values of the wavelength of the beam diffracted by the sample. If only a number of discrete wavelengths is required, this can be done by changing the orientation of the monochromator so that reflexion takes place at a different family of planes. If a continuous variation of wavelength is needed, the glancing angle

θ_M of the neutrons striking the reflecting planes must be capable of continuous alteration. In this second case, the bulky shielding around the monochromator must be designed to allow extraction of the beam at a variable scattering angle $2\theta_M$. It must be noted, however, that as θ_M changes there is not only a continuous change in the wavelength λ_M but also a variation in the wavelength spread $\delta\lambda$ (p. 207) and in the focusing position of the beam diffracted by the sample (p. 208).

TABLE XVIII. *Single crystal neutron monochromators*

Crystal	Special features	References
Lead	Tends to have 'lineage' structure, causing non-uniformity of reflected beam; low Debye–Waller factor e^{-2W}, and so unsuitable for high scattering angles $2\theta_M$	Shull & Wollan, 1951; Alikhanov, 1959
Copper	Beam more uniform in cross-section than for lead; good reflectivity, even at high $2\theta_M$	Shull & Wollan, 1951
Germanium	No $\frac{1}{2}\lambda$ contamination, if (111) is reflecting plane; large unit cell and so suitable for long wavelengths; poor reflectivity in as-grown condition	McReynolds, 1952; Shull, 1960; Barrett, Mueller & Heaton, 1963
Beryllium	Low mosaic spread, but high intrinsic reflectivity; small absorption so that thick crystals can be used in transmission; Debye–Waller factor near to unity and so suitable for high $2\theta_M$	—
LiF, NaCl, CaF$_2$	Tend to be more perfect than metal crystals, and to have lower intrinsic reflectivities Q	Sturm, 1947; Wollan & Shull, 1948
Magnetite	Suitable for long wavelengths	McReynolds, 1952

Harmonic contamination of primary beam striking sample

A crystal monochromator, set to reflect the wavelength λ_M from the (hkl) family of planes, will also simultaneously reflect neutrons of wavelengths λ_M/n (where $n = 2, 3, ...$) from the $n(hkl)$ planes at the same crystal orientation. Harmonics of wavelengths $m\lambda_M$ and $(m/n)\lambda_M$ would also be reflected if h, k, l have a common factor m such that the corresponding plane gives an allowed reflexion.

For $\lambda_M \approx 1$ Å the flux is still appreciable at $2\lambda_M$, $3\lambda_M$, ... (Fig. 85), and so the reflecting plane must be chosen with this second possibility in mind. Thus the (422) plane of copper can be used, as the 211 reflexion is absent, but not the (222) plane. For the λ_M/n harmonics, we calculate from equation (7.5) that the flux at $\lambda = \frac{1}{2}\lambda_m$ is only 2 per cent of that at $\lambda = \lambda_m$ (peak of wavelength distribution curve), and the flux at $\frac{1}{3}\lambda_m$, $\frac{1}{4}\lambda_m$, ... is smaller still. The weak 'second-order contamination', $\frac{1}{2}\lambda_M$, may constitute a serious handicap to the examination of weak Bragg reflexions: in studies of magnetic structures, for instance, it is often necessary to distinguish between a weak magnetic reflexion, obtained by reflecting the wavelength λ_M, and a nuclear reflexion occurring at the same angular position and corresponding to the diffraction of the wavelength $\frac{1}{2}\lambda_M$ by a strongly reflecting plane with half the spacing.

Several methods are available for reducing or eliminating this second-order contamination. By orienting the monochromator to scatter neutrons at wavelengths lower than the peak wavelength λ_m, some intensity is sacrificed but the second-order contamination is reduced appreciably (see Fig. 87). A more effective method is to use a filter. For example, ^{239}Pu has a sharp resonance absorption for slow neutrons of energy 0·295 eV, corresponding to a wavelength of 0·53 Å: by selecting a monochromatic beam of 1·06 Å and passing the beam through a thin layer of plutonium, the $\frac{1}{2}\lambda_M$ component is attenuated. Suitable thicknesses of plutonium filter for reducing the second-order contamination by a given amount have been calculated by Atoji (1964); Atoji also tabulates the characteristics of other resonance filters which can be used in the wavelength range 0·7–1·8 Å. In germanium single-crystal monochromators, the second-order component is suppressed automatically by choosing a reflecting plane such as (111) with a forbidden second-order reflexion. Finally, the second-order and higher-order components can be removed by placing a mechanical velocity selector in the incident beam (see p. 214).

Width of reflected wavelength band

The neutron monochromator isolates a narrow band of wavelengths from the wavelength spectrum emerging from the collimator. The ratio of the wavelength spread $\delta\lambda$ to the mean wavelength λ_M is

$$\frac{\delta\lambda}{\lambda_M} = \cot\theta_M \, \delta\theta_M, \qquad (7.6)$$

where θ_M is the Bragg angle of the monochromator and $\delta\theta_M$ is the angular spread of the reflected beam. If the monochromator is a

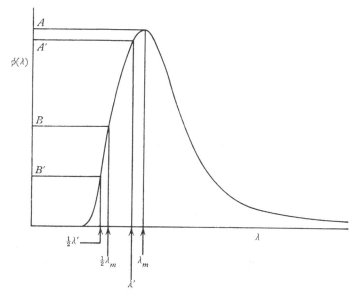

Fig. 87. The second-order contamination $\phi_{B'}/\phi_{A'}$ is less at a wavelength λ' (where $\lambda' < \lambda_m$) than the contamination ϕ_B/ϕ_A at the peak wavelength λ_m.

perfect crystal, the angular spread $\delta\theta_M$ arises solely from the divergence of the beam emerging from the collimator. For a collimation angle of $\pm\frac{1}{2}°$, $\delta\theta_M$ is $1\cdot0° = 0\cdot017$; from (7.6)

$$\frac{\delta\lambda}{\lambda_M} = 1\cdot7 \text{ per cent} \quad \text{for} \quad \theta_M = 45°,$$

and $\qquad \dfrac{\delta\lambda}{\lambda_M} = 10 \text{ per cent} \quad \text{for} \quad \theta_M = 10°.$

An additional contribution to $\delta\theta_M$ arises from the divergence produced by the mosaic spread of the monochromator. However, we can estimate roughly that for $\lambda_M \sim 1$ Å the wavelength band has a width of 0·02 Å for $\theta_M = 45°$ and a width greater than 0·10 Å for $\theta_M = 10°$.

At high values of θ_M the bandwidth is less, and so the resolution of the Bragg reflexions is improved. However, the overall intensity of the primary beam falls, first, because of the narrower wavelength band selected by the monochromator, and second, because of the effect of the Debye–Waller factor in reducing the reflectivity of the monochromator. Another effect, however, tends to outweigh the loss of intensity in measuring reflexions at high values of θ_M: this is the focusing effect, which is well-known in the theory of the double-crystal X-ray spectrometer (Compton & Allison, 1935).

Focusing effect

Let us suppose that the collimator is of such a length and width that it permits passage of neutrons making angles $\pm \alpha$ to the central path, and that the neutrons in this central path are scattered at a glancing angle θ_M by the monochromator. The glancing angles for the extreme paths of the neutrons in the horizontal plane are then $\theta_M + \alpha$ and $\theta_M - \alpha$ (Fig. 88). In the vertical plane the range of Bragg angles is much less, $\theta_M \pm \frac{1}{2}\tan\theta_M \alpha^2$, and so we need discuss only neutrons in the horizontal plane. We shall ignore the effect of the mosaic spread of the monochromator in increasing the divergence of the primary monochromatized beam, although the mosaic spread is incorporated readily in the more general treatment (Willis, 1960).

If scattering at a glancing angle θ_M corresponds to a wavelength λ_M, then scattering at $\theta_M + \alpha$ corresponds to a wavelength

$$\lambda_M(1 + \alpha\cot\theta_M),$$

and scattering at $\theta_M - \alpha$ to

$$\lambda_M(1 - \alpha\cot\theta_M).$$

(These formulae follow from equation (7.6).) Thus the monochromator not only reflects a narrow band of wavelengths, but it sorts out in angle the wavelengths within this band: the shorter wavelengths emerge at the side of low scattering angles and the longer wavelengths at high angles.

The integrated intensities for the various *hkl* reflexions of a single crystal sample are now measured by placing the sample in this monochromatic beam and oscillating the crystal through the reflecting range. If the scattered intensity is plotted against ω, where ω is the angle of rotation about the goniometer-head axis, the area under this curve gives the integrated intensity. This intensity curve can be obtained with the sample scattering the

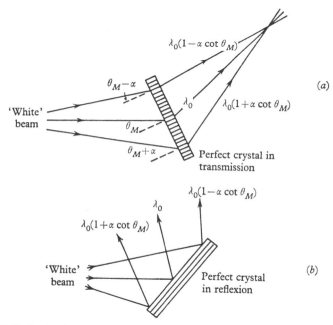

Fig. 88. Reflexion introduces a correlation between wavelength and direction of scattered beam. The correlation occurs whether the monochromator operates in transmission (*a*) or in reflexion (*b*).

monochromatic beam either to the right (Fig. 89*a*) or to the left (Fig. 89*b*): for a given reflecting plane the intensity integrated over the full reflecting curve is the same in the two cases, but the widths of the reflecting curves are quite different. The reason for this difference is that the various wavelengths in the 'monochromatic' beam from the crystal monochromator have been sorted out in angle: for scattering to the right the longer wavelengths strike the second crystal at larger glancing angles to the reflecting plane, while the longer wavelengths are incident at smaller glancing angles in scattering to the left. Thus in Fig. 89 a the full range of glancing angles for the band of wavelengths $\delta\lambda$ is covered in a smaller rotation of ω than in Fig. 89 b, that is, the reflecting curves are narrower in (a). In the special case for which the reflecting

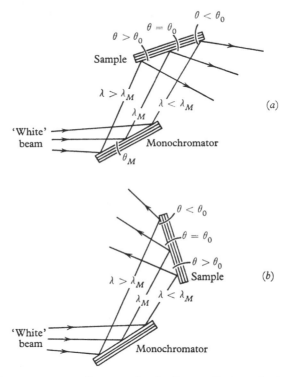

Fig. 89. Two arrangements for measuring the Bragg reflexions of a single crystal sample. The 'parallel' arrangement (a) is used in preference to (b), because the longer wavelengths in (a) have larger glancing angles whereas the reverse is true in (b): this difference gives rise to sharper reflexions from the sample in (a).

planes of the specimen and monochromator are parallel, all wavelengths are reflected to the right at the same angular position of the specimen: the width of the reflecting curve is then independent of the collimation angle of the primary beam and is determined solely by the mosaic spreads of the monochromator and the specimen.

To obtain narrow reflecting curves and hence good signal-to-background ratios from the sample the 'parallel' arrangement of Fig. 89a is used. The theory of the dependence of rocking-curve width on Bragg angle θ of the sample has been derived for the parallel arrangement by a number of authors (Willis, 1960; Dachs, 1961; Caglioti & Ricci, 1962). The full-width at half height H of the reflexion occurring at a Bragg angle equal to the glancing angle of the monochromator ($\theta = \theta_M$) is

$$H = 2\sqrt{[\ln 2(\eta^2 + \eta_M^2)]},$$

where η is the mosaic spread of the sample and η_M the mosaic spread of the monochromator. ('Mosaic spread' is defined on p. 174.) For other values of θ, H depends on the collimator angle α as well as on η and η_M. The nature of this dependence is shown in Fig. 90, where H is plotted against the quantity $\tan\theta/\tan\theta_M$, the 'resolution parameter'. The divergence of the beam passing down the primary collimator is $\pm\alpha$, and each pair of broken and unbroken curves in Fig. 90 refers to a different value of α: the unbroken curve in each pair corresponds to $\eta = \eta_M$ and the broken curve to $\eta = 0$.

The width of the reflecting curve is a minimum in the region of $\theta = \theta_M$, and θ_M should be chosen to make the width about the same at either end of the θ range under investigation. This implies that, for a range of 0–60° in θ, θ_M is about 40°. Moreover, by using relatively high values of θ_M, the hkl reflexions from the sample are made sharp at high Bragg angles, where the intensities are reduced more by thermal vibration.

The advantages in using a high glancing angle θ_M are demonstrated by the experimental data shown in Fig. 91. The reflecting curves were recorded using a single crystal of ThO_2 as the specimen, and the fall-off of peak intensity with scattering angle 2θ is very marked for measurements taken with a small monochromator angle, $\theta_M = 11°$. The 755 reflexion, occurring at $2\theta = 134°$, is easily observed using $\theta_M = 45°$, but for $\theta_M = 11°$ it is so broad that it has almost disappeared into the general background. We see then that, although there is a fall in the intensity of the incident beam arising from a reduction in its wavelength range, increasing the glancing angle θ_M gives a striking improvement in the signal-to-background ratio. For the measurement of weak reflexions an improvement in the signal-to-background ratio is more important than increasing the signal alone (see p. 274).

In the study of magnetic structures the maximum value of 2θ for the sample is generally less than $50°$ for an incident wavelength in the neighbourhood of 1 Å. This is so because magnetic reflexions arise from the scattering of neutrons by the outer unpaired electrons

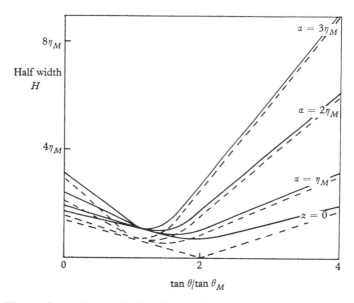

Fig. 90. Dependence of half-width of reflecting curve on resolution parameter $\tan\theta/\tan\theta_M$. Full curves for sample with same mosaic spread as monochromator ($\eta = \eta_M$); broken curves for sample consisting of perfect crystal ($\eta = 0$). $\pm\alpha$ is the divergence of the beam striking the monochromator (Willis, 1960).

of the atom, so that there is a steep fall-off in magnetic scattering amplitude with θ. The form-factor dependence of magnetic scattering contrasts with nuclear scattering, which is independent of θ because the scattering centres are nuclei whose dimensions are smaller by a factor of 10^{-4} than the wavelength of thermal neutrons. Thus for magnetic work the most suitable value of θ_M, or the value which gives focusing conditions near the middle of the accessible 2θ range, will be considerably lower than for general studies involving the nuclear reflexions alone.

Uniformity of monochromatized beam

In measuring a set of reflexions from a single crystal it is usual to bathe the crystal completely in the incident beam and to assume that each reflecting plane intercepts the same proportion of the beam while the crystal is turned from one reflexion to the next. This is not true, however, if the crystal is not spherical and if the incident monochromatized beam is not uniform in cross-section.

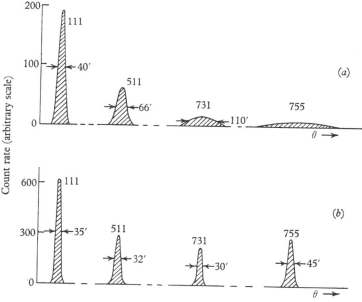

Fig. 91. Experimental reflecting curves for ThO_2 single crystal: (a) $\theta_M = 11°$, (b) $\theta_M = 45°$. The focusing position is near the 111 reflexion in (a) and near 731 in (b). A different collimator was used for recording the two sets of measurements (a) and (b) (Willis, 1962b).

The uniformity of the beam used in neutron diffractometry is typically ± 5 per cent over an area of 3×3 mm at the specimen. Fig. 92 shows the contours of constant neutron flux in the beam reflected by a single-crystal copper monochromator: the measurements were made by scanning the beam with a $\frac{1}{2}$ mm pinhole of cadmium placed at the position of the specimen and recording the transmitted beam with a BF_3 detector.

For the highest accuracy in measuring the relative values of the integrated intensities, spherical specimens are used. Errors will still arise from a lack of uniformity of the incident beam, if the absorption of the specimen is large. Fortunately, in most neutron studies the absorption is very small.

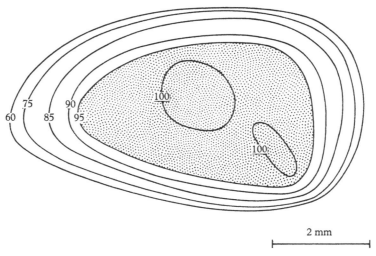

2 mm

Fig. 92. Experimental contours of equal neutron flux in a beam reflected by a copper-crystal monochromator. The beam is uniform in intensity to ± 5 per cent in dotted region. Contours marked on an arbitrary scale.

7.4. Mechanical monochromators

Neutron diffractometers are normally operated in association with crystal monochromators, but there seems to be no fundamental reason why mechanical neutron-velocity selectors, or mechanical monochromators, should not be used instead. Moreover, a mechanical selector is more versatile in that it allows a choice in the values of both λ and $\delta\lambda/\lambda$ within fairly wide limits. Mechanical selectors are particularly attractive for wavelengths greater than 2 Å, as the second-order and higher-order harmonics, appearing well up the wavelength distribution curve (Fig. 85), would be very troublesome with a crystal monochromator.

The flight time between the specimen and the detector for the neutrons used in neutron-beam experiments is between 10^{-5} and 10^{-3} sec, and it is possible, therefore, to use mechanical choppers suitably spaced apart to produce monochromatic neutrons of selected λ and $\delta\lambda$. The principle of the mechanical selector for slow neutrons is illustrated in Fig. 93. Several discs, containing cadmium to make them opaque to neutrons, are mounted on the same axis, which is parallel to the neutron beam. Each disc contains a radial slit which transmits neutrons as the disc rotates. The continuous flux of reactor neutrons is chopped by the first disc into a neutron pulse having a broad wavelength spectrum. This burst then travels a known distance to the second disc, whose slit has a fixed phase

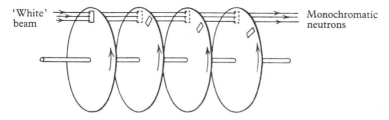

'White' beam   Monochromatic neutrons

Fig. 93. Principle of mechanical velocity selector for slow neutrons (after Hughes, 1953).

relation with respect to the first. When it opens, the second disc passes only a band of wavelengths, and both the mean wavelength and the width of the band are related to the time-of-flight of the neutrons between the two discs, the relative phase of the discs and the open-time of the slits. Greater flexibility in varying both λ and $\delta\lambda$ is achieved by using a multi-rotor system of three or four rotating discs.

Curved slot rotors are used in the apparatus for cold neutron measurements described by Otnes & Palevsky (1963). Here the path of the neutrons is normal to the axis of rotation and the neutrons pass down a curved slot in the rotor (Fig. 94) to emerge in a broad wavelength band. The slot follows a circular arc, which is a sufficiently good approximation to the ideal curve (an Archimedean spiral). The values of λ and $\delta\lambda/\lambda$ can be varied independently between 3·5 and 10 Å and between 1·5 and 6 per cent respectively; this variation is achieved by using several rotors, which are operated at various speeds and separations.

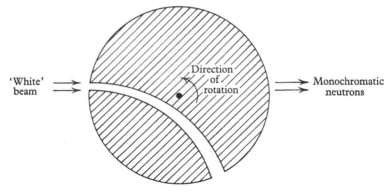

Fig. 94. Curved slot rotor.

7.5. Neutron diffractometry using 'white' primary beam

Slow neutrons have a speed of about 4,000 m/s at a wavelength of 1 Å, and this speed can be readily measured by time-of-flight techniques. Moreover, the speed varies inversely as the wavelength of the neutrons, in accordance with equation (7.3), and this gives rise to the possibility of carrying out experiments in neutron diffractometry using a radically different technique from any discussed so far.

If the direct beam from the nuclear reactor is pulsed with a mechanical chopper, then all those neutrons recorded at a given instant of time after the pulse have the same wavelength. The chopper with a single electronic time-gate constitutes a device for producing and recording monochromatic neutrons: if there are many adjacent time channels, diffraction of a range of wavelengths can be investigated simultaneously.

A single-channel arrangement for elastic scattering studies with long wavelength neutrons (~ 5 Å) has been developed by Low & Collins (1963). A diagram of their apparatus is shown in Fig. 95. The direct neutron beam passes through a filter of polycrystalline beryllium, which removes by Bragg scattering all those neutrons with a wavelength less than 3·95 Å, representing twice the maximum lattice spacing in beryllium. The beam then passes through a filter of single-crystal bismuth, which serves to attenuate the gamma radiation. Thus the final emergent beam consists predominantly of slow neutrons with wavelengths greater than 4 Å. A simple chopper pulses the beam and the neutrons scattered by the specimen are recorded in BF_3 counters after traversing a 1 m flight path. The counter assembly is gated in synchronism with the chopper, with a gating delay corresponding to the time-of-flight of neutrons with a wavelength of about 5 Å. The mean wavelength and the wavelength spread are changed by altering the gating delay and gate width, respectively.

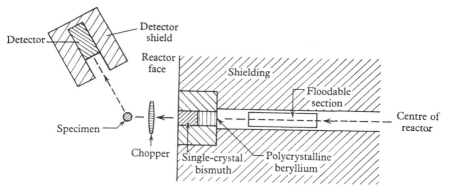

Fig. 95. Apparatus for elastic scattering studies with 5 Å neutrons (after Low & Collins, 1963).

The second type of experimental arrangement, using many adjacent time channels, has been developed recently by Buras & Leciejewicz (1964) as a modified form of the Laue method. (A similar technique was proposed by Lowde (1956), who emphasized that higher counting rates are achieved by integrating the diffracted intensity over wavelength rather than crystal angle.)

In the conventional method of crystal structure analysis, neutrons of a fixed wavelength are scattered by the sample and the intensity $I(\theta)$ of the diffracted beam is measured as a function of the variable angle θ (Fig. 96a). The curve of $I(\theta)$ versus θ shows a maximum when the Bragg equation, $\lambda = 2d\sin\theta$, is satisfied by the particular set of planes with spacing d. However, it is possible to reverse the roles of λ and θ, and to measure the intensity $I(\lambda)$ as a function of the wavelength λ at a fixed value of the scattering angle 2θ (Fig. 96b). A single disc of the type used in Fig. 95 chops the slow neutron beam from the reactor, giving a pulse of neutrons with widely varying wavelengths at the specimen; the sample then scatters those neutrons with wavelengths satisfying the Bragg equation, and the scattered neutrons are counted by means of a neutron detector connected to a multichannel time-analyser. The wavelength is determined from the distance between the specimen and detector and the time taken to traverse this distance.

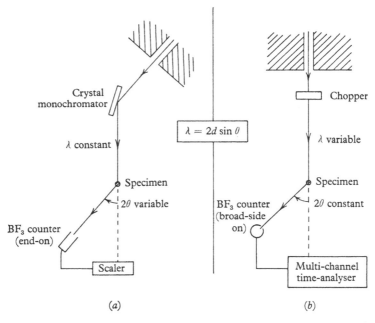

Fig. 96. (a) Conventional method, and (b) time-of-flight method of measuring Bragg reflexions. To define the flight-path precisely the BF_3 counter is used in the 'broadside-on' position in (b). (After Buras & Leciejewicz, 1964.)

Fig. 97 shows a neutron diffraction pattern of nickel powder taken by the time-of-flight method (Schwartz, 1965, private communication). Distinct peaks appear, which are readily indexed from a knowledge of the wavelength at which they occur. So far the method has been used for powder specimens only, but it could be applied equally well to the examination of single crystals. It is particularly promising for neutron diffraction work at long wavelengths as the counting rate is proportional to λ^4.

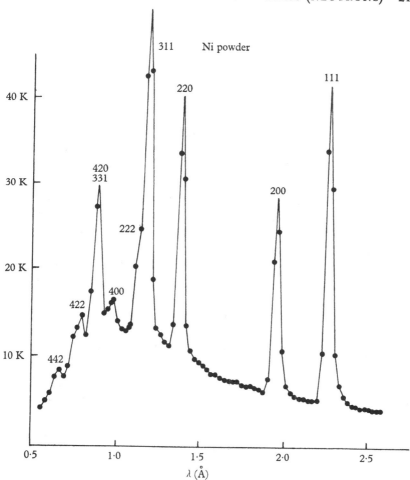

Fig. 97. Neutron diffraction pattern of nickel powder obtained by
time-of-flight method (Schwartz, 1965).

CHAPTER 8

THE BACKGROUND

8.1. Introduction

Before describing the measurement of the coherent Bragg reflexions, we shall discuss the various sources of background scattering occurring with the reflexions. The background scattering introduces both systematic and random errors in the determination of the diffracted intensity. Systematic errors arise from those components of the background scattering which have a non-linear dependence on scattering angle in the neighbourhood of the Bragg peak: the contribution of the background to the peak cannot then be estimated by simply extrapolating background measurements taken on either side of the peak. The random or statistical error associated with the presence of the background is discussed later in Chapter 10: we shall note there the importance of reducing the background as much as possible in order to improve the statistics of counting, especially in measuring weak reflexions.

Thus the conditions under which the reflexions are measured must be chosen with two points in mind: first, we must obtain a good estimate of the background under the Bragg peak, in order to make a valid background subtraction; secondly, the background must be as small as possible, so as to enhance the signal-to-background ratio.

The principal contributions to the background are the following:

(1) Contributions which peak at the Bragg reflexions.

    (a) Bragg scattering of harmonics or subharmonics of the fundamental wavelength (p. 205). These must be removed by one of the techniques discussed in Chapters 6 and 7, since corrections are not normally possible.

    (b) Thermal diffuse scattering (§8.2).

(2) White radiation background: X-rays only (§8.3).

(3) Contributions which vary slowly with scattering angle.

(a) Incoherent-scattering (§§8.4 and 8.5) in the form of

Fluorescence scattering (X-rays)

Compton scattering (X-rays)

Isotope incoherent scattering (neutrons)

Spin incoherent scattering (neutrons)

(b) Parasitic scattering from amorphous regions in the specimen, from the specimen support, and from the air surrounding the specimen (mainly with X-rays) (§8.6).

(4) Detector background: in X-ray counters due to cosmic radiation and radioactive contamination; in neutron counters due to γ-rays and fast neutrons (see Chapter 4).

We shall now examine some of these contributions.

8.2. Thermal diffuse scattering

Shortly after the first experiments in X-ray diffraction, Debye (1914) gave a mathematical treatment of the effect of thermal motion on the scattering of X-rays by a crystal. His calculations showed that thermal motion causes a decrease in the intensity, but not in the sharpness, of the Bragg reflexions by a factor e^{-2W}, where

$$W = 8\pi^2 \overline{u_s^2} \frac{\sin^2\theta}{\lambda^2}. \qquad (8.1)$$

In this expression $\overline{u_s^2}$ is the mean-square displacement of the atoms along the normal to the reflecting planes. In addition, Debye demonstrated that an equivalent intensity should appear as a general diffuse background.

Faxén (1923) showed that this diffuse background ('thermal diffuse scattering', or TDS) was not uniform throughout reciprocal space, but possessed broad maxima which were centred at the reciprocal lattice points. The extra spots observed by Laval (1938) on Laue photographs were identified as those predicted by Faxén and by Waller (1928) as due to thermal motion. Laval's work stimulated a wide interest in the thermal diffuse scattering of X-rays by matter: the study of X-ray diffuse scattering is now a well-established technique in lattice dynamics, giving information about vibrational frequency spectra, the dispersion curves for elastic waves propagating along the principal crystallographic directions, and the atomic force constants of solids. Neutron inelastic scattering has been used with even greater success in recent years to derive the same kind of information.

Nilsson (1957) first discussed the effect of thermal diffuse scattering on the measurement of the Bragg reflexions and showed that the diffuse scattering can make a considerable contribution to the measured intensity under the Bragg peak. The scattering rises to a maximum under the peak, and so a large part of the TDS is included if we assume, in the usual way, that the diffracted intensity is represented by the peak intensity minus the background intensity at the sides of the peak. In the case of the high-angle reflexions of rock salt observed with CuKα radiation, Nilsson showed that the integrated intensity derived in this way is over-estimated by as much as 30 per cent.

The amount of thermal diffuse scattering which is included in the measurement of a Bragg reflexion increases with the volume of reciprocal space illuminated during the intensity scan. Burbank (1964) has calculated this volume for the ω and $\omega/2\theta$ scans: his recommendations for minimizing the illuminated volume under different conditions are listed on p. 267.

The diffuse scattering under the Bragg peak has been observed directly by O'Connor & Butt (1963) who used 14·4 keV (0·86 Å) X-rays from a Mössbauer source. The diffuse scattering of X-rays in the region of the Bragg peaks involves the exchange of energy between the X-rays and the longest-wavelength phonons in the crystal; typically, the change of energy might be 10^{-9} eV. The Mössbauer effect permits the separation of the radiation scattered elastically (without energy change) from the inelastically scattered component. Fig. 98 shows some recent results obtained by O'Connor & Butt on the reflexion of 14·4 keV X-rays by the (111) plane of a single crystal of aluminium and by the (200) plane of a single crystal of KCl. It is seen that the inelastic intensity does indeed increase at the Bragg angle, as predicted theoretically. The extremely weak intensity of Mössbauer sources limits their general use for separating the elastic and inelastic components of the Bragg reflexions.

Chipman & Paskin (1958) have corrected for the TDS effect in X-ray powder work by measuring the wings of the Bragg peak

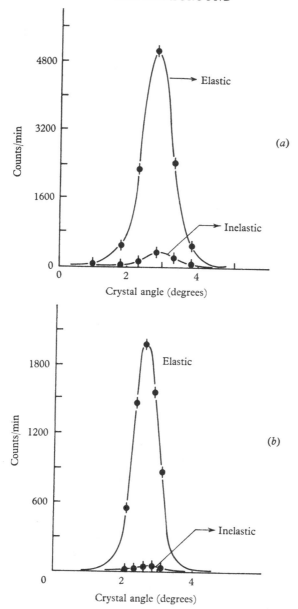

Fig. 98. Angular variation of elastic and inelastic scattering in the neighbourhood of: (a) 111 reflexion from aluminium; (b) 200 reflexion from potassium chloride (after O'Connor & Butt, 1965).

where the scattering is contributed by TDS alone, and calculating from these measurements the TDS contribution at the centre of the Bragg peak. This method has not been used in single crystal work: the calculation would require a knowledge of the spectrum

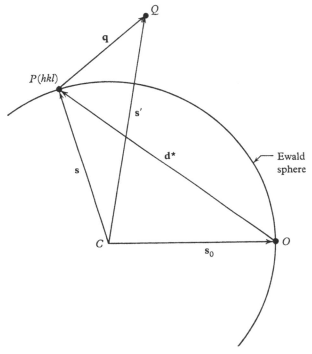

Fig. 99. Geometry of single-phonon annihilation process. **q** is the wave-vector of the phonon absorbed by the crystal and **d*** is the reciprocal lattice vector of the *hkl* point. The elastically scattered neutrons **s** can be separated from the inelastic neutrons **s'**, provided $|\mathbf{s'}|$ is very different from $|\mathbf{s}|$ ($= |\mathbf{s_0}|$). In the single-phonon creation process, the end-point of **q** lies inside the Ewald sphere and $|\mathbf{s'}|$ is less than $|\mathbf{s_0}|$.

of thermal vibrations of the atoms in the crystal and of the principal elastic constants of the crystal—information which is certainly not available for the majority of crystals.

Caglioti (1964) has examined the possibility of eliminating the inelastic component of the Bragg peak in neutron diffraction. The energy of a single phonon is comparable with that of a thermal neutron, and so the neutron suffers an appreciable change of energy when it is scattered and emits, or absorbs, one phonon. The situation is depicted in Fig. 99, which shows the geometry of the single-phonon annihilation process. A neutron of wave vector $\mathbf{s}_0$ is

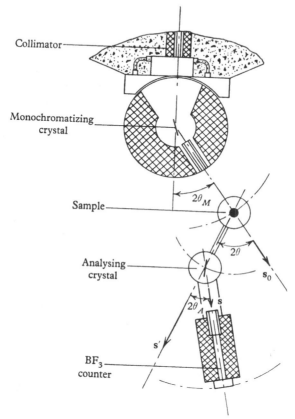

Fig. 100. A triple-axis spectrometer adapted for 'elastic' neutron diffraction. As in a conventional diffractometer, the monochromator provides a beam of neutrons of wavelength λ_0 impinging on the sample. The angle $2\theta_A$ of the analysing crystal is set to reflect neutrons of wavelength λ_0. In principle, only those diffracted neutrons of energy corresponding to λ_0 are allowed to be Bragg-reflected by the analyser and to reach the BF_3 detector. In practice, the energy resolution of the spectrometer limits the degree of discrimination between elastically and inelastically scattered neutrons (after Caglioti, 1964).

scattered in the direction $\mathbf{s}'$ by the single-phonon interaction, and, if the magnitude of $\mathbf{s}'$ $(= s')$ is very different from unity $(s_0 = 1)$, the inelastically scattered neutron can be prevented from reaching the BF_3 detector by using a 'triple-axis spectrometer' of the type illustrated in Fig. 100. The analysing crystal of the spectrometer is set for Bragg reflexion of the neutrons having the same energy as those impinging on the specimen, and neutrons of very different energy are not Bragg reflected.

The ability to discriminate in this way between inelastically and elastically scattered neutrons depends critically on the energy resolution of the analysing spectrometer. This resolution is governed by the angular divergences of the neutrons passing between the sample and analyser and between the analyser and detector; to maintain a reasonable neutron intensity these divergences must be about 30' of arc, and this angle gives an energy resolution of a few meV (10^{-3} eV). Neutrons interacting with phonons of lower energy than this cannot be separated from the elastically scattered neutrons, and so energy selection is poor near the reciprocal lattice points where the diffuse intensity is high and is contributed by low-energy single-phonon processes ($\sim 10^{-9}$ eV). For those substances which could be studied most efficiently by the 'elastic' diffraction method, namely 'hard' materials characterized by comparatively high values of the Debye temperature, the intensity of the thermal diffuse scattering is intrinsically low anyway. Thus, because the use of an analysing crystal causes a loss of intensity of the diffracted beam, there is probably little gain in employing such a crystal to reduce the effect of thermal diffuse scattering in the neighbourhood of the neutron reflexions.

There are, in fact, no generally satisfactory experimental techniques, either for X-rays or for neutrons, which lead to the elimination or correction of the TDS contribution to the Bragg reflexion. However, the ratio of the true integrated intensity arising from elastic scattering, to the intensity which is measured on the assumption of a linear background under the Bragg reflexion, is of the approximate form $e^{-2W'}$, where

$$W' = \text{constant} \frac{\sin^2\theta}{\lambda^2} \qquad (8.2)$$

(Nilsson, 1957). This means that W' is of the same form as the W factor in equation (8.1). Thus W' and W can be combined and the effect of thermal diffuse scattering treated as an artificial increase in the Debye–Waller factor: in many crystallographic studies the temperature factors are considered as empirical quantities whose magnitudes do not seriously affect the final results of the investigation. It is only in certain investigations, such as the determination of the characteristic Debye temperature from the experimental Debye–Waller factors, that serious difficulties of interpretation are likely to arise.

8.3. White radiation background

An X-ray beam which is not monochromatic gives rise to a 'white radiation background'. As the crystal is rocked through the Bragg reflecting position, different wavelengths in the continuous spectrum which is superimposed on the characteristic radiation are Bragg-reflected into the counter. The white radiation background is not uniform across the reflexion, and its form depends on the type of intensity scan adopted in measuring the reflexion.

Let us consider the idealized case of the diffraction of a parallel beam of strictly monochromatic X-rays by a very small and perfect crystal. The reciprocal lattice point $P(hkl)$ is brought onto the surface of the sphere of reflexion by turning the crystal through an angle ω_0: the corresponding diffracted beam emerges in the direction $\mathbf{s} = \overrightarrow{CP}$ (see Fig. 101 a) at an angle $2\theta_0$ to the incident beam. If the detector is kept stationary at $2\theta_0$ and the crystal oscillated through an angular range from $\omega_0 - \delta\omega_0$ to $\omega_0 + \delta\omega_0$, the detector effectively moves along a circular path in reciprocal space with centre at the origin O. This path corresponds to the 'ω-scan', in which the crystal moves while the detector is stationary. On the other hand, in the $\omega/2\theta$-scan the crystal and detector are co-ordinated in a $1:2$ angular motion, and the detector moves through reciprocal space along a line passing radially through O.

If the X-ray beam contains a white radiation component, wavelengths other than the primary monochromatic wavelength

$$\lambda_0 \quad (= 2d_{hkl}\sin\theta_0)$$

are diffracted by the (hkl) plane along a radial line in reciprocal space, as shown in Fig. 101 b. The magnitude of the reciprocal lattice vector is λ/d, so that the point P slides radially along this line for different wavelengths λ. The finite size and mosaic spread of the crystal cause diffraction to occur along a narrow cone rather than a line in reciprocal space.

According to Fig. 101, the intensity contributed to the background of the Bragg reflexion by the white radiation differs in the

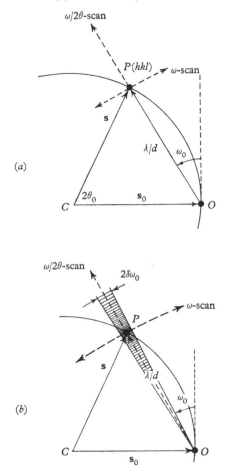

Fig. 101. (a) Idealized measurement of *hkl* reflexion by ω-scan and by $\omega/2\theta$-scan; (b) real case showing effects of white radiation streak and of finite size and mosaic spread of crystal (after Alexander & Smith, 1962).

ω-scan and $\omega/2\theta$-scan. In the ω-scan the path of the scan is tangential to the radial reciprocal lattice vector and proceeds across the white streak (Fig. 101 b) from points in reciprocal space lying well outside the 'ridge' of white radiation. The resultant experimental profile (see Fig. 102) shows a Bragg peak superimposed on a low background, but the peak contains a large contribution from wavelengths other than λ_0 in the white radiation zone. In contrast to this, the path of the $\omega/2\theta$-scan lies along the radial vector, that is, along the ridge of the white radiation zone: the background is

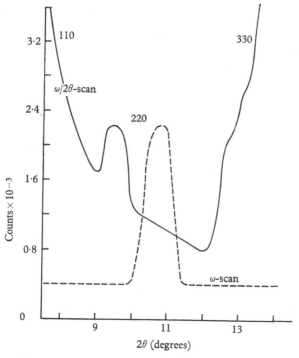

Fig. 102. Experimental profiles of 220 reflexion of $(C_6H_5)_4$Sn obtained with $\omega/2\theta$-scan (full curve) and ω-scan (broken curve): unmonochromatized MoKα radiation (after Alexander & Smith, 1962).

higher on either side of the reflexion than for the ω-scan, but the extrapolated background under the peak is closer to the true background.

The argument, given above, for interpreting the ω-scan and the $\omega/2\theta$-scan as tangential and radial motions respectively of the detector through reciprocal space, assumes that the aperture or window of the detector is infinitesimally small. Burbank (1964) points out that the correct interpretation of the two types of scan requires consideration of the finite size of the detector aperture and, in fact, by choosing the aperture appropriately for the two procedures the profile of the reflexion and the white radiation background can be made the same for either scan. The Laue streak will merely be shorter with the ω-scan and will appear as a pair of shoulders instead of a continuous background. Burbank shows that, if the aperture for the ω-scan is made wide enough, a line drawn through the shoulders will allow an adequate subtraction of the white radiation background.

Fig. 102 also illustrates another point. In the ω-scan the detector remains at a fixed position and will not catch radiation diffracted by neighbouring reciprocal lattice points. For this reason, resolution of adjacent reflexions is better with the ω-scan than with the $\omega/2\theta$-scan.

With monochromatized radiation the white radiation background is absent, but other difficulties are introduced by using monochromators in X-ray work (p. 188).

In neutron diffraction there is no white radiation background, as the beam is monochromatized.

8.4. Incoherent scattering of X-rays

The scattering of X-rays due to the Compton effect and the photoelectric effect (fluorescence scattering) is incoherent, as the wavelength of the scattered radiation is modified and there is no fixed phase relationship between the wavelets of the scattered beam. The absence of interference effects means that incoherent scattering is distributed fairly evenly throughout reciprocal space: it does not peak at the Bragg reflexions and so does not cause difficulties in deriving the true integrated intensity of the type associated with the coherent thermal diffuse scattering and the coherent white radiation background. This type of background, therefore, affects the statistics of counting only.

The fluorescence yield for any element is strongly dependent on the wavelength of the incident radiation: it is greatest when this wavelength is a little shorter than a prominent absorption edge of one of the elements in the specimen. Thus it is almost impossible to use CuKα radiation ($\lambda = 1\cdot54$ Å) with a specimen containing a large proportion of cobalt (K absorption edge at 1·61 Å) or iron (K absorption edge at 1·74 Å), and a longer wavelength radiation must be used. However, fluorescence is also caused by the white radiation in the primary beam and this is present whatever the choice of X-ray tube target. The effects of specimen fluorescence can often be greatly reduced by inserting a suitable filter between the sample and the detector: this filter must have an absorption edge a little longer than the characteristic radiation excited in the specimen.

In Compton scattering the incident X-ray quantum loses energy by collision with one of the outer electrons of an atom in the sample (see, for example, Compton & Allison, 1935). According to classical theory, the increase in wavelength for a scattering angle 2θ is given by

$$\Delta\lambda = \frac{h}{mc}(1 - \cos 2\theta), \qquad (8.3)$$

where $h/mc = 0\cdot024$ Å, and so the maximum increase in wavelength is 0·05 Å. This is too small to allow any significant reduction of the Compton component by filtration. Fortunately, the intensity of the Compton scattering is always small compared with that of the Bragg scattering.

8.5. Incoherent scattering of neutrons

In neutron diffraction there are two main sources of incoherent scattering. 'Isotope incoherence' is produced if the sample contains a random distribution of isotopes, each of which has a different scattering amplitude. Incoherent scattering also occurs when the neutrons are scattered by atoms of non-zero nuclear spin: this 'spin incoherence' is especially prominent with hydrogen-containing samples but it can be appreciably reduced by replacing hydrogen by deuterium. There is no incoherent scattering with samples which contain single isotopes with zero nuclear spin.

The two neutron incoherent scattering components are both unmodified in wavelength and are examples of disorder scattering. This is in contrast with the incoherent X-ray contributions of §8.4 which are of a different wavelength from the elastically scattered radiation.

Apart from hydrogen spin incoherence, neutron incoherent scattering is usually too small to be very troublesome.

8.6. Parasitic scattering

Radiation may be scattered by materials surrounding the specimen, especially if this is enclosed in a high- or low-temperature container; biological crystals frequently need to be in contact with the mother liquor of crystallization. Precautions consist of reducing the amount of such foreign matter to a minimum and selecting the material of the container or of the mounting cement with regard to its scattering and fluorescence properties.

X-rays are also scattered by the column of air which surrounds the specimen: this volume can be reduced by the use of a backstop and a detector collimator (p. 172).

8.7. Background and rocking range

Equation (6.1) is an expression for the minimum angle through which the crystal rotates to allow a valid measurement to be made of the integrated intensity. Generally, it is not safe to employ a rocking range which approaches this minimum value: lack of precision in the value of the unit cell dimensions, back-lash in the shafts of the diffractometer, and setting errors all combine to provide a small uncertainty in the peak position. The crystal must be oscillated through an angle rather bigger than the minimum indicated by equation (6.1) to make certain that the complete reflexion is measured. The 'peak-to-background' ratio deteriorates in proportion to the unnecessary background which is taken in on either side of the reflexion: it is very important that the observed and calculated peak positions coincide as closely as possible, if the peak-to-background ratio is poor and so requires the rocking range to assume the smallest permissible value. The safest procedure in such cases is to use some method of hunting for the peak. This can be done in a number of ways. A very large number of intensity ordinates are recorded across the reflexion: when the diffractometer output is processed, the computer program inspects the slope of the intensity curve and determines the points at which the reflexion profile rises above the background. This procedure is practicable only with a diffractometer which is connected on-line to a computer: the volume of output data becomes unmanageable on punched cards or punched tape. An on-line computer offers the alternative possibility of keeping the actual scanning range small and of locating the peak precisely before the measurement is made. It is not necessary to hunt for the peak of every reflexion: instead a smaller number of reflexions strategically distributed in reciprocal space can be selected so that reflexions in their neighbourhood will then be located with sufficient accuracy. Special-purpose peak-hunting circuits have been incorporated in the off-line diffracto-meter 'Cascade' (Cowan, Macintyre & Werkema, 1963).

CHAPTER 9

SYSTEMATIC ERRORS IN MEASURING
RELATIVE INTEGRATED INTENSITIES

In a crystal structure determination the measured intensity of the reflexions must be converted to a set of relative 'integrated intensities'; from these a set of structure factors is derived by applying various geometrical and scale factors (see Chapter 11). The integrated intensity, also called 'integrated reflexion', is defined in the following way.

Let the crystal be placed in an incident beam, which has a uniform intensity I_0 and bathes the crystal completely. The crystal is rotated with uniform angular velocity ω through the reflecting position for the (hkl) family of planes: the angle of rotation is sufficient to allow all the mosaic blocks in the crystal to diffract the complete band of wavelengths $\delta\lambda$ comprising the 'monochromatic' beam. If E is the total amount of energy diffracted from the incident beam, then the integrated intensity ρ_{hkl} is:

$$\rho_{hkl} = E\omega/I_0. \qquad (9.1)$$

Before the quantities ρ_{hkl} defined by (9.1) can be converted to structure factors, which express the amplitude of scattering in the hkl direction from a single unit cell, we must relate the integrated intensities for the macroscopic crystal to the integrated intensity for an infinitesimal volume element δV. This requires an examination of various physical factors, such as absorption, extinction and simultaneous reflexions (Renninger effect), which affect the magnitudes of the observed intensities from a macroscopic crystal. It is customary to consider these factors to be systematic errors in measuring the integrated intensities. Systematic errors associated with thermal diffuse scattering and with the white X-radiation background have been described in the previous chapter.

9.1. Absorption

X-rays and neutrons are absorbed as well as scattered in their passage through matter, although the relative importance of the two processes is markedly different for the two radiations. Thus the absorption of X-rays due to the photo-electric effect is much larger than the attenuation due to scattering, whereas in neutron diffraction scattering is a principal cause of attenuation of the neutron beam. To relate the intensity scattered by a single atomic plane to that scattered by the whole crystal, we must know the amount by which both the incident and diffracted beams are reduced by absorption in the crystal. This calculation must be carried out separately for each reflexion and depends on the path of the incident and diffracted beams through the crystal.

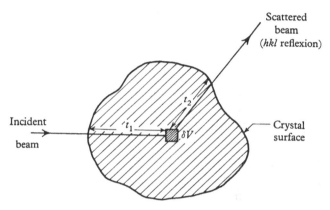

Fig. 103. Scattering from volume element δV of crystal.

The amount by which the intensity of the hkl reflexion is reduced by absorption is denoted A_{hkl}. The reciprocal of A_{hkl} is the absorption factor $A^*_{hkl} = 1/A_{hkl}$: A^* represents the factor by which the observed intensity is multiplied to obtain the corrected intensity.

Fig. 103 shows the volume element δV scattering the incident beam into the hkl reflexion. t_1 is the path length of the incident beam inside the crystal before being scattered by δV, and t_2 is the path length of the diffracted beam. The absorption factor is

$$A^*_{hkl} = \frac{V}{\displaystyle\int_V e^{-\mu x}\,dV}, \qquad (9.2)$$

where μ is the linear absorption coefficient of the crystal for the particular type of incident radiation used, $x = t_1 + t_2$, and V is the volume of the crystal.

The linear absorption coefficient

If a narrow beam of monochromatic radiation passes through a thickness t of the crystal, the emergent intensity I is related to the incident intensity I_0 by

$$I = I_0 e^{-\mu t}. \tag{9.3}$$

(We assume the crystal is oriented to avoid a Bragg reflexion.)

In X-ray diffraction the dominant contribution to μ is the true absorption arising from the photoelectric effect (fluorescence). μ is computed from the relation

$$\mu = \rho_c \left\{ p_1 \left(\frac{\mu}{\rho}\right)_1 + p_2 \left(\frac{\mu}{\rho}\right)_2 + p_3 \left(\frac{\mu}{\rho}\right)_3 + \dots \right\}, \tag{9.4}$$

where ρ_c is the density of the crystal, p_1, p_2, ... are the fractions by weight of elements 1, 2, ... in the crystal, and $(\mu/\rho)_1$, $(\mu/\rho)_2$, ... their mass absorption coefficients. *The International Tables for X-ray Crystallography*, volume III, contains tables of the mass absorption coefficients of the elements.

Values of μ for typical single crystals and MoKα and CuKα radiations are listed in Table XIX, taken from a paper of Jeffery & Rose (1964).

TABLE XIX. *Values of μ for typical specimens examined with MoKα and CuKα radiations*

μ (cm^{-1})

MoKα	CuKα	Specimen
1·5	10	Molecular organic compounds with no atoms heavier than oxygen
4	30	Sodium salts of organic acids; inorganic crystals composed of atoms of low atomic number and containing large amounts of water of crystallization
12–80	100	Organic bromides; inorganic crystals composed of light elements, and heavier elements with water of crystallization
35–200	300	Organic crystals containing fairly heavy atoms; inorganic crystals composed of elements in the intermediate range of atomic numbers
200	500	Crystals containing very heavy elements

In neutron diffraction, a few elements (including lithium, boron and cadmium) have relatively high values of μ, but for the majority of elements μ is less than $1 \cdot 0$ cm^{-1}. Values of μ for typical specimens examined with 1 Å neutrons are given in Table XX. Incoherent scattering processes, particularly nuclear spin incoherence from nuclei with non-zero spin, may contribute more than true absorption to the effective absorption coefficient μ. Thus for the hydrogen nucleus there is a difference in sign, as well as in magnitude, of the scattering amplitudes for the two spin states of the compound nucleus formed between the neutron and proton: this difference gives rise to appreciable incoherent scattering, which is isotropic and accounts for the relatively large μ value of 2–5 cm^{-1} in Table XX for hydrated compounds. μ can be computed for a particular crystal using tables of nuclear cross-sections, but it is preferable to determine the effective coefficient experimentally by measuring the attenuation of a narrow monochromatic beam in passing through a known thickness of the sample.

TABLE XX. *Values of μ for typical specimens examined with neutrons of wavelength* $1–1 \cdot 5$ Å

μ (cm^{-1})	Specimen
2–5	Organic or inorganic crystals containing large amounts of water of crystallization
10–100	Crystals containing lithium, boron or cadmium
0·01–0·5	Nearly all other crystals not containing H, Li, B, Cd

Absorption factor for spherical crystal

For $\theta = 0°$ and $\theta = 90°$, equation (9.2) can be integrated directly for a sphere of radius R. The result is

$$A^* = \tfrac{2}{3}(\mu R)^3 \left[\tfrac{1}{2} - e^{-2\mu R}(\tfrac{1}{2} + \mu R + \mu^2 R^2)\right]^{-1}$$

for $\theta = 0°$, and

$$A^* = \frac{4}{3}\mu R \left[\frac{1}{2} - \frac{1}{16\mu^2 R^2}(1 - e^{-4\mu R} - 4\mu R e^{-4\mu R})\right]^{-1}$$

for $\theta = 90°$. For other values of θ, A^* can be evaluated by numerical methods; A^* is tabulated at $5°$ intervals between 0 and $90°$ and for $\mu R = 0$, 0·1, 0·2, ..., 10·0 in *The International Tables for X-ray Crystallography*, volume II. The same volume includes a similar table for cylindrical crystals, examined in two dimensions, with the incident and scattered beams normal to the cylinder axis. The correction of the observed intensities for absorption in the sample is most conveniently applied, in both X-ray and neutron diffraction, when the crystal is ground into a spherical (or cylindrical) shape.

TABLE XXI. *Absorption factor A* for a spherical crystal*

μR	$\theta = 0°$	$15°$	$30°$	$45°$	$60°$	$75°$	$90°$
0·0	1·00	1·00	1·00	1·00	1·00	1·00	1·00
0·1	1·16	1·16	1·16	1·16	1·16	1·16	1·16
0·2	1·35	1·34	1·34	1·34	1·33	1·33	1·33
0·3	1·56	1·55	1·55	1·53	1·52	1·51	1·51
0·4	1·80	1·79	1·78	1·75	1·73	1·70	1·70
0·5	2·08	2·06	2·03	1·99	1·94	1·91	1·90
0·8	3·15	3·11	2·99	2·83	2·69	2·58	2·55
1·0	4·12	4·03	3·79	3·50	3·25	3·07	3·01
1·5	7·80	7·38	6·44	5·52	4·83	4·39	4·23
2·0	14·0	12·6	10·0	7·96	6·59	5·78	5·50
2·5	23·8	20·0	14·5	10·7	8·46	7·23	6·80
3·0	38·4	29·9	19·5	13·6	10·4	8·70	8·11
4·0	86·5	56·0	31·0	19·8	14·4	11·7	10·8
5·0	167	89·1	43·5	26·4	18·5	14·7	13·4
6·0	288	127	56·8	33·0	22·7	17·8	16·1
8·0	683	213	84·4	46·6	31·1	23·9	21·4
10·0	1333	308	113	60·5	39·0	30·1	26·7

Table XXI, which is a section from the *International Tables* giving the absorption correction for a spherical crystal, brings out a number of important points. The figures in each vertical column show that $A*$ varies rapidly with μR, particularly at low Bragg angles, and so slight departures from the ideal spherical shape cause an appreciable uncertainty in the estimate of the absorption correction. Jeffery & Rose (1964) have calculated the magnitude of this uncertainty, expressed as the fractional standard deviation $\sigma(A*)/A*$, as a function of θ. They assumed an average variation in the radius, $\sigma(R)/R$, of 2·5 per cent, corresponding to the average variation of the spheres used in their investigation. Their results, plotted in Fig. 104, show that small departures from sphericity can introduce large errors in the determination of the integrated intensities, and for this reason alone the relative integrated intensities of many X-ray samples cannot be measured to an accuracy better than a few per cent.

Reading the figures across in Table XXI we note that there is no need to apply the absorption correction for a perfect sphere, provided that μR is very small ($<$0·2) and provided that relative intensities are required only. If, however, the intensities are to be measured on an absolute scale, the absorption correction must be applied even at very low values of μR: for $\mu R =$ 0·1, the correction

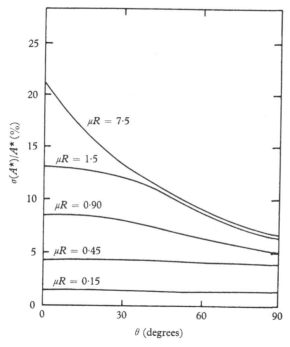

Fig. 104. Graphs showing uncertainty in absorption correction for a spherical crystal due to a 2·5 per cent uncertainty ($\sigma(R)/R$) in the radius (after Jeffery & Rose, 1964).

to the absolute intensities is already as high as 16 per cent. The greater significance of the absorption correction in absolute intensity measurements is one reason why these are more difficult to make than relative intensity measurements (p. 290).

For a cylindrical crystal or a flat plate examined in transmission, absorption causes a fall in the total diffracted energy above a certain crystal size, but for a spherical crystal there is no such critical size and the energy increases continuously from $R =$ 0 to

$R = \infty$. This is illustrated by Fig. 105 in which the quantity R^3/A^* is plotted against μR: R^3 is proportional to the crystal volume and R^3/A^* to the total diffracted energy.

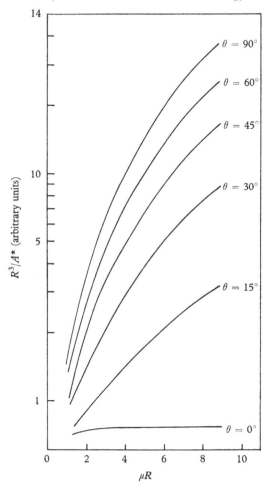

Fig. 105. Total diffracted intensity for spherical crystal plotted against μR. There is no 'critical size' at which the total intensity is a maximum.

If possible, the crystal should be sufficiently small for the absorption correction to vary only slowly with θ and large enough to give an adequate intensity. From these criteria the optimum size would correspond to a μR value of about 1·0.

Absorption factor for any crystal

The correction of single crystal intensity measurements for the effect of absorption is greatly simplified if the crystal is ground into a sphere or cylinder, but often this is not possible. Grinding the crystal may lead to fragmentation or twinning; shaping is also difficult if there are strong cleavage planes.

For an unground crystal, if the dimensions of the faces bounding the crystal are measured, A^* can be evaluated either graphically or with a computer. In Albrecht's graphical method (Albrecht, 1939), the specimen is divided into small volume elements and the integral $\int_V e^{-\mu x} dV$ in equation (9.2) evaluated by summing $e^{-\mu x}$ for each element. Other graphical methods are described in *The International Tables for X-ray Crystallography*, volume II: they are all very laborious and their accuracy decreases as the magnitude of the correction increases.

Computer programs for deriving the absorption corrections for single crystal specimens of arbitrary shape bounded by flat faces have been described by Busing & Levy (1957) and by Wells (1960). Wells also considers specimens mounted in a tube in which an absorbing liquid may be trapped between the crystal and the tube wall: biological crystals are often mounted in this way. The calculation of the absorption correction with these programs is relatively slow and may require as much as one second per reflexion even on a large computer such as the IBM 7090.

Another approach to the absorption problem has been suggested by Furnas (1957). The method makes use of the fact that the azimuthal orientation ψ of the reflecting plane of the crystal can be changed, thereby altering the total radiation path $x = t_1 + t_2$, without destroying the Bragg reflecting condition. Both extinction and simultaneous reflexions give rise to variations in the intensity with ψ (see §§9.2 and 9.3), and so both these effects must be small for the method to work satisfactorily.

An empirical method based on this idea has been developed by North, Phillips & Mathews (1966). A low-angle reflexion, whose scattering vector is parallel to the ϕ-axis, is observed with a four-circle diffractometer; the azimuth ψ is altered simply by changing

ϕ and so the intensity can be plotted as a function of azimuth (Fig. 106c). In Fig. 106a, AA and BB are the projections of the incident and reflected beams for two particular values of ϕ, and the

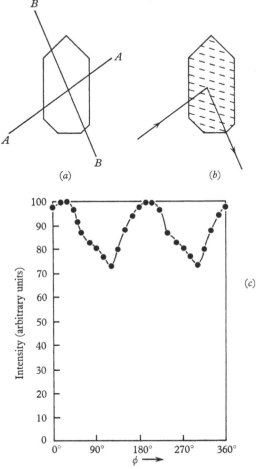

Fig. 106. Absorption correction using the method of North, Phillips & Mathews. (a) Reflecting plane in plane of paper and ϕ-axis normal to paper; (b) reflecting planes normal to plane of paper and incident (reflected) beams parallel to AA (BB); (c) azimuthal scan for reflexion in (a) showing effect of absorption.

absorption correction for a reflexion (Fig. 106b) incident along AA and reflected along BB is taken as the mean of the absorptions along the paths AA and BB in setting (a). The method is most effective with low-angle reflexions where it is possible to derive the absorption for the projected paths to a close approximation; with protein crystals absorption corrections of up to 40 per cent, determined by this method, have been shown to agree closely with those determined analytically by the much slower method of Busing & Levy (1957).

The absorption factor A^*, given by the expression (9.2), is often the most serious source of error in X-ray work. This is so because A^* cannot be evaluated precisely, unless the crystal has a well-defined shape and size or has a low linear coefficient of absorption μ. In neutron diffraction the μ values are normally much lower: consequently, the absorption effect is small and an adequate correction can be applied relatively easily.

9.2. Extinction

Extinction is the attenuation of the incident beam which is caused by Bragg reflexion. If the crystal is perfect, the apparent absorption along directions in which Bragg reflexions occur may be many times as large as the ordinary value: the incident beam can only penetrate a short distance into the crystal before being reflected, and the inner parts of the crystal have no chance to diffract the radiation. This enhanced absorption in a crystal with a perfect arrangement of atoms is called 'primary extinction'.

Most crystals, however, possess irregularities in their atomic arrangement, in the form of dislocations, point defects, sub-grain boundaries, and the like. These irregularities tend to destroy the coherence, or fixed phase relationship, between the components of the incident beam scattered by different parts of the crystal. The crystal is divided effectively into small regions, perhaps 10,000 Å across, which are sufficiently perfect to reflect the beam coherently, but between which there is no coherence. These perfect regions are known as 'mosaic blocks', and if there is negligible primary extinction of the beam in passing through a single block the crystal is 'ideally mosaic'.

The passage of an X-ray or neutron beam through a mosaic crystal will be quite different from that through a perfect crystal. The angular range of reflexion for a perfect crystal is only a few *seconds* of arc, whereas the angular misorientation of adjacent mosaic blocks is measured in *minutes* of arc. Consequently, the beam will penetrate deeply into a mosaic crystal before it reaches mosaic blocks which are identical in orientation to those near the surface and which will reflect the same part of the beam again. Attenuation of the beam by Bragg scattering from identically oriented mosaic blocks is known as 'secondary extinction' (see Fig. 107). In primary extinction the *amplitudes* of the scattered waves must be added to obtain the scattering from the whole crystal; for secondary extinction there is no coherence between the mosaic blocks and the *intensities* must be summed.

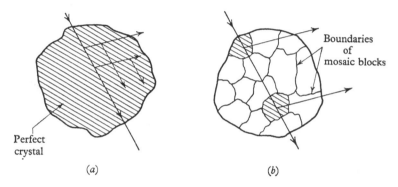

(a)   (b)

Fig. 107. (a) Primary extinction, causing attenuation of beam in a perfect crystal, or in a single mosaic block. (b) Secondary extinction, causing attenuation by reflexion at two mosaic blocks with same orientation.

In spite of the large volume of theoretical work on extinction, the correction factors to be applied to the measured integrated intensities for taking into account the affect of extinction can be calculated exactly in a few idealized cases only. Thus Zachariasen (1945), Bacon & Lowde (1948) and James (1962) have based their theoretical treatments of extinction on crystals in the form of infinite plane-parallel plates, whereas we are usually concerned with finite crystals which are completely bathed in the beam. An approximate treatment of secondary extinction for crystals of any shape has been given by Hamilton (1957, 1963), and this will be discussed later. Our main purpose here, however, is to describe the procedure of detecting and correcting for extinction, and we shall only sketch the relevant theoretical background. The reader interested in the detailed theory of extinction is recommended to study the books by Zachariasen (1945) and James (1962).

Both primary and secondary extinction are dependent on the strength of the reflexion, on the wavelength of the incident beam and on the dimensions of the crystal. A number of methods for detecting and correcting for extinction exploit these various relationships.

Dependence on intensity

The effect of extinction is most pronounced for the strongest reflexions, because these scatter the maximum amount of energy

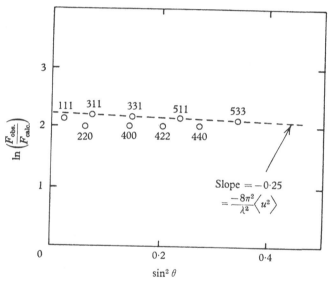

Fig. 108. Secondary extinction in UO_2. The intensities of the strong reflexions, with even indices, are reduced by extinction (after Willis, 1963).

from the incident beam. One procedure in structure determinations is to ignore the strongest reflexions in the initial stages of the least-squares refinement of the data, in which the observed structure factors ($F_{obs.}$) are matched with those calculated from the trial structure; they may then be introduced in the later refinement stages, and any systematic tendency for $F_{obs.}$ to be lower than $F_{calc.}$ indicates the presence of extinction. An example of the detection of extinction in this way is illustrated by Fig. 108, showing results obtained by neutron diffraction in the examination of a single crystal of UO_2. The quantity $\ln(F_{obs.}/F_{calc.})$ is plotted against $\sin^2\theta$, and as $F_{calc.}$ was derived without taking into account the effect of thermal motion this curve should be a straight line with a slope of $-(8\pi^2/\lambda^2)\langle u^2 \rangle$, where $\langle u^2 \rangle$ is the mean square displacement of the atoms due to thermal agitation. The reflexions with odd indices have points lying on this straight line, but the stronger reflexions with indices obeying the relation $h+k+l = 4n$ are contributed by atoms which are all scattering in-phase, and the corresponding points lie below the line because of extinction.

Chandrasekhar (1956, 1960a, b) has devised an elegant method of correcting for extinction in X-ray diffraction, which does not require any previous knowledge of the atomic positions in the unit cell and which applies to any kind of extinction, primary, or secondary, or both. In the absence of extinction, the intensity reflected by a mosaic crystal is proportional to the square of the structure factor

$$I_{obs.} = \alpha|F|^2,$$

and Chandrasekhar has shown that to a first approximation this formula is modified to

$$I_{obs.} = \alpha|F|^2 - \beta|F|^4 \tag{9.5}$$

by the effects of extinction. α is a known factor whose calculation is described in Chapter 11; β is unknown and depends on the nature and amount of extinction. If the intensity is measured with plane-polarized X-rays, the diffracted intensities will be different for the two states of polarization in which the plane of polarization is perpendicular or parallel to the plane of incidence. Equation (9.5) can be applied to the two states separately, giving

$$I_{obs.}^{\perp} = \alpha|F|^2 - \beta|F|^4 \tag{9.6}$$

and $$I_{obs.}^{\parallel} = \alpha|F|^2\cos^2 2\theta - \beta|F|^4\cos^4 2\theta. \tag{9.7}$$

$I_{obs.}^{\perp}$ and $I_{obs.}^{\parallel}$ denote the intensities which are measured with the X-rays polarized in the perpendicular and parallel directions. The structure factors for the two polarization states differ by the factor $\sin(90°-2\theta)$ (see p. 284), and this difference accounts for the replacement of F in (9.6) by $F\cos 2\theta$ in (9.7). Eliminating the unknown β from these two equations leads to

$$|F|^2 = \frac{1}{\alpha} \frac{I^{\parallel}_{\text{obs.}} - I^{\perp}_{\text{obs.}}\cos^4 2\theta}{(\cos^2 2\theta - \cos^4 2\theta)}. \qquad (9.8)$$

Thus by making two measurements for each reflexion, using X-rays polarized perpendicular and parallel to the plane of incidence, an extinction-free estimate of $|F|^2$ is derived by applying equation (9.8).

The principal limitation of Chandrasekhar's method is that equation (9.8) is ill-conditioned for values of θ approaching $0°$, $45°$ and $90°$: the extent of this limitation depends on the accuracy with which the intensities are measured. The incident beam can be polarized by one of the two methods discussed on p. 192. The loss of intensity associated with the polarization of the beam by either of these methods, and the experimental difficulties of repeating each measurement under identical conditions for each polarization state, have prevented the wide adoption of Chandrasekhar's method.

A modified form of the method has been used in neutron diffraction for the study of magnetic reflexions (Chandrasekhar & Weiss, 1957; Szabo, 1961). However, it cannot be used in general neutron work as there is no polarization effect for nuclear reflexions.

Dependence on wavelength

The amount of extinction decreases with decreasing wavelength, and we can write approximately (Chandrasekhar, 1960*b*):

$$I_{\text{obs.}} = I_{\text{corr.}}\left[1 - \left(P^2 + \frac{S^2}{\mu}\right)p\lambda^2\right]. \qquad (9.9)$$

Here $I_{\text{obs.}}$ is the observed intensity and $I_{\text{corr.}}$ the intensity corrected for the effect of extinction; p is the polarization factor, λ the wavelength and P, S are constants representing the amount of primary, secondary extinction respectively. A wavelength dependence of the relative values of $I_{\text{obs.}}$ for different reflexions indicates the presence of extinction. The linear absorption coefficient μ for X-rays varies with λ^3 between absorption edges, so that $I_{\text{obs.}}$ is not a linear function of λ^2 unless there is no secondary extinction ($S = 0$ in equation 9.9).

Dependence on crystal size

To a first approximation, the effect of secondary extinction is to increase the absorption coefficient from μ to $\mu + gQ$, where g is the 'secondary extinction coefficient' (related to the mosaic spread of the crystal) and Q is the quantity:

$$Q = N_c^2 \lambda^3 (e^2/mc^2)^2 Lp F^2 \tag{9.10}$$

(Darwin, 1922). N_c is the number of unit cells per unit volume and Lp is the Lorentz-polarization factor (see Chapter 11): the term $(e^2/mc^2)^2$ is replaced by unity in neutron diffraction (Hamilton, 1957). Thus the attenuation of the beam inside the crystal is $e^{-(\mu+gQ)x}$, where x is the optical path length; if the secondary extinction is small ($gQx \ll 1$), the attenuation is

$$(1 - gQx)e^{-\mu x}.$$

The term $e^{-\mu x}$ represents the effect of true absorption and $1 - gQx$ is the factor by which the observed intensity decreases on account of secondary extinction.

Thus the effect of secondary extinction can be eliminated by measuring the intensity, corrected for true absorption, for different values of x, and extrapolating to zero path length. The procedure was first used in X-ray diffraction by Bragg, James & Bosanquet (1921), and more recently by Cochran (1953) and by Witte & Wölfel (1955). Cochran measured the relative intensities of the reflexions from spherical crystals of different radii, and corrected for extinction by plotting the changes in relative intensities against the radius of the spheres.

The method of varying path length has also been used in neutron diffraction (Willis, 1962c). A convenient method of altering the path length is to change the azimuthal angle ψ. Figure 109 shows the neutron intensities of the $h$oo reflexions of KBr, measured at different azimuthal orientations ψ of the (100) plane. The crystal was in the form of a flat plate, cleaved parallel to the {100} faces; for $\psi = 0°$ or 180° the total path length x was a minimum, and the path was longest for $\psi = 90°$. The experimental measurements were corrected for the small effect of true absorption: they showed that the intensity of the weakest reflexion 800 was independent of the path length through the crystal, but for 600, 400, 200 the maximum intensity occurred at the minimum path length. The

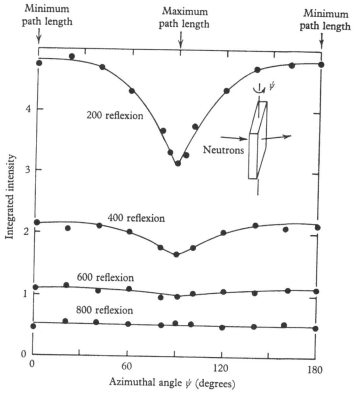

Fig. 109. Integrated intensities of $h$oo reflexions of KBr crystal as a function of ψ (after Willis, 1962 c).

proportional reduction in intensity due to secondary extinction increased with the strength of the reflexion and was largest for the strongest reflexion, 200. The slight fluctuation of the experimental points from the smooth curves in Fig. 109 may be due to the influence of simultaneous reflexions (see §9.3). The derivation of the corrected intensities from these curves is described in the original paper.

Hamilton (1957, 1963) has discussed the calculation of secondary extinction corrections for crystals of any arbitrary polyhedral shape. If the extinction is large, so that the intensities are reduced by 30 per cent or more, Hamilton recommends that the crystal be ground to a cylindrical shape, and he has prepared tables for the correction of the equatorial reflexions of cylindrical crystals. A section of these tables is reproduced in Table XXII: the numbers represent the factors by which the observed intensities are multiplied to correct them for secondary extinction, and are given as a function of 2θ and the quantity aRQ. R is the radius of the cylinder, Q is given by (9.10) and is proportional to the intensity of the reflexion, and a is a numerical constant related to the mosaic spread of the specimen. The tables cannot be used directly as a is normally unknown, but Hamilton has written a computer program, based on these tables, which allows a least-squares comparison of the observed and calculated intensities: a is treated as an extra empirical constant to be determined in the least-squares procedure.

TABLE XXII. *Secondary extinction corrections for cylindrical crystal (Hamilton 1963)*

aRQ	$\theta = 0°$	$22·5°$	$45°$	$67·5°$	$90°$
0·0	1·000	1·000	1·000	1·000	1·000
0·2	1·178	1·177	1·173	1·170	1·169
0·4	1·374	1·368	1·354	1·341	1·336
0·6	1·585	1·571	1·538	1·511	1·501
0·8	1·811	1·785	1·727	1·682	1·666
1·0	2·050	2·007	1·917	1·852	1·830
2·0	3·389	3·196	2·880	2·700	2·641
3·0	4·863	4·437	3·846	3·545	3·444
4·0	6·393	5·684	4·812	4·387	4·143
6·0	9·490	8·172	6·741	6·070	5·832
8·0	12·61	10·65	8·667	7·752	7·416
10·0	15·75	13·12	10·59	9·432	8·996

Hamilton assumes that the effective absorption coefficient in the presence of secondary extinction is $\mu + gQ$, in accordance with the original treatment of Darwin (1922). However, Zachariasen (1963) has shown that, for incident unpolarized X-radiation, Darwin's formula is in error, and that the effective coefficient is instead

$$\mu + 2gQ \frac{(1 + \cos^4 2\theta)}{(1 + \cos^2 2\theta)^2}.$$

This expression reduces to $\mu+gQ$ at $\theta = 0°$ and $\theta = 90°$ but to $\mu+2gQ$ at $\theta = 45°$. The error arises from the incorrect treatment of the effect of polarization in Darwin's theory; the theory remains strictly valid for neutrons, because there is no polarization of nuclear scattering. Thus Table XXII applies to the scattering of neutrons from a cylindrical specimen, but the numbers in the columns for $\theta = 22\cdot5°$, $45°$, $67\cdot5°$ require small corrections for the scattering of unpolarized X-rays.

None of the methods for correcting for the effect of extinction is entirely satisfactory; they are all based on theoretical approximations and several involve an elaborate experimental technique, which may not be justified by the final results. If the extinction correction is small, requiring, say, a 5 per cent correction to the measured intensities, a reasonable correction is possible, but for large extinction a rough correction only can be made.

The safest procedure for dealing with extinction is to ignore those reflexions for which extinction is suspected. This is legitimate in the refinement of intensity data, when atomic co-ordinates are to be determined by least-squares methods: it is not permissible in deriving electron density distributions, where all reflexions must be included up to the maximum observed Bragg angle. It may be possible to reduce extinction by grinding the crystal or by subjecting it to thermal shock, but this will rarely eliminate extinction altogether (Lonsdale, 1947).

9.3. Simultaneous reflexions

In interpreting the intensity reflected by a family of planes (hkl) it is normally assumed that no other Bragg reflexion is occurring at the same time as the hkl reflexion under observation. However, at certain crystal orientations the Bragg reflecting condition may be satisfied for more than one family of planes, and the presence of a second reflexion can modify the intensity of the first. The occurrence of simultaneous reflexions is by no means uncommon, and may even be unavoidable, particularly when working at short wavelengths or with crystals of large inter-planar spacings. Fortunately, the magnitudes of the intensity fluctuations caused by simultaneous reflexions tend to be relatively small in X-ray diffraction, and the effect has received little attention in the X-ray literature since the work of Renninger (1937), who was the first to make a thorough study of simultaneous reflexions. More recently, with the advent of counter methods and the demand for higher accuracy, further X-ray studies have been reported (Cole, Chambers & Dunn, 1962; Cohen, Fraenkel & Kalman, 1963; Zachariasen, 1965). In neutron diffraction, simultaneous reflexions are more troublesome, because the specimens are larger and absorb neutrons less readily than X-rays: consequently, the neutron beam penetrates a long way into the crystal, and there is a greater chance of energy being exchanged between different families of reflecting planes. Several authors have examined experimentally the influence of simultaneous reflexions on the measurement of integrated intensities with neutrons (Borgonovi & Caglioti, 1962; Willis, 1963; Moon & Shull, 1964).

The geometrical problem of determining those crystal orientations for which two or more reflexions occur is conveniently treated in terms of the reciprocal lattice and the Ewald sphere of reflexion. The condition for a single reflexion to occur is that the Ewald sphere passes through the hkl reciprocal lattice point, and this condition is maintained if the sphere rotates about the axis OP passing through the hkl point P and the origin of the reciprocal lattice O (see Fig. 110a). During this rotation the sphere sweeps through the reciprocal lattice and each time it touches a second reciprocal lattice point, such as $h'k'l'$, the conditions are satisfied for hkl and $h'k'l'$ to occur simultaneously.

In Fig. 110a the directions of the incident beam and the first and second reflexions are denoted by the unit vectors s_0, s and s'. The beam s_0 is scattered along the directions s, s', but the beams s and s' can then each be rescattered along two directions, giving rise to six scattering processes in all. The rescattering processes for the first reflexion s are illustrated by Fig. 110b. The once-scattered beam in the direction s is treated as the 'incident beam' for the second scattering process and we can transfer the origin of the reciprocal lattice to the end-point of s. The process $s \rightarrow s_0$ takes place by rescattering of the first reflexion at the family of planes $(\bar{h}\bar{k}\bar{l})$, and $s \rightarrow s'$ by rescattering at the family with indices $(h' - h, k' - k, l' - l)$. Both processes cause a diminution of intensity of the first reflexion s. An intensity reduction is also contributed by the process $s_0 \to s'$, which removes power from the incident beam and thereby decreases the power available for the reflexion $s_0 \to s$, whereas a gain in intensity follows the transfer of energy to the first reflexion by the roundabout process ('Umweganregung'):

$$s_0 \to s' \to s.$$

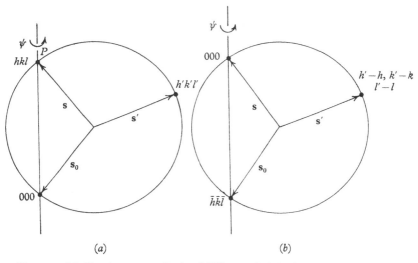

(a)   (b)

Fig. 110. (a) Simultaneous reflexion $h'k'l'$ recorded during rotation of crystal about hkl scattering vector $\overrightarrow{OP}$. The circle represents a section of the Ewald sphere through OP. (b) Once reflected hkl beam in (a) can be rescattered by planes with indices $\bar{h}\bar{k}\bar{l}$ and $h'-h$, $k'-k$, $l'-l$.

Clearly, the net effect on the intensity of the hkl (or s) reflexion is very difficult to calculate and depends on the reflecting powers of the planes $\pm(hkl)$, $\pm(h'k'l')$ and $\pm(h-h', k-k', l-l')$. If more than two reciprocal lattice points lie simultaneously on the Ewald sphere, the number of different scattering processes is much larger.

Experimental curves showing the intensity in the direction of the primary reflexion are shown as a function of ψ in Figs. 111 and 112. Fig. 111 shows a large number of intensity peaks at different values of ψ when a germanium crystal is so oriented that the 222 reciprocal lattice point is on the Ewald sphere. These peaks occur even though the 222 reflexion is a forbidden one in the diamond structure of germanium. The extra peaks are caused by the round-about excitation of 222 by the process $s_0 \to s' \to 222$: s_0 is the incident beam and s' is a reflexion which occurs simultaneously with 222. In all, there are over two hundred peaks in a 360° rotation of ψ, and these are associated with a corresponding number of simultaneous reflexions s'. Similarly, the true 200 reflexion in the neutron diffraction pattern of iron (Fig. 112) is the base line between the peaks (and dips) caused by simultaneous reflexions: at $\lambda = 1.57$ Å there are seven calculated positions of ψ in a 45° range at which reflexions occur simultaneously with 200, while at $\lambda = 0.72$ Å this number increases to seventy-eight.

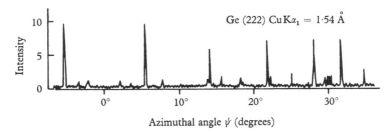

Fig. 111. Simultaneous reflexions observed with X-rays in the 'forbidden' 222 reflexion from germanium, as the crystal is rotated about the 222 scattering vector (after Cole, Chambers & Dunn, 1962).

An approximate theory of Moon & Shull (1964) accounts for the intensity changes caused by simultaneous reflexions. It applies to crystals in the shape of flat plates which are larger than the incident beam cross-section and is restricted to small secondary extinction and low absorption.

In the absence of a general theory allowing the calculation of the correction to the observed intensities, the most satisfactory practical procedure is to take steps either to avoid or to minimize the effect of simultaneous reflexions. An obvious precaution is to choose the azimuthal angle ψ of the reflecting plane hkl (the angle of rotation about the line OP in Fig. 110a), so that the Ewald sphere is as far as possible from all other reciprocal lattice points. The calculation of the 'forbidden' ψ values which give rise to simultaneous reflexions is straightforward: a graphical method for cubic crystals is described by Cole, Chambers & Dunn (1962), and an analytical method, suitable for crystals of any symmetry, by Santoro & Zocchi (1964). Powell (1966) has written a computer program which derives suitable setting angles of a four-circle diffractometer for minimizing simultaneous reflexion effects; for a particular reflexion there is an infinite number of combinations of setting angles, each of which corresponds to a different value of ψ,

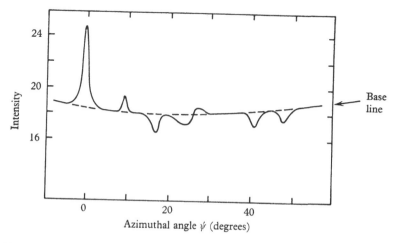

Fig. 112. Simultaneous reflexions observed with neutrons ($\lambda = 1\cdot57$ Å) in the 200 reflexion from iron, as the crystal is rotated about the 200 scattering vector (after Moon & Shull, 1964).

and the particular combination selected is that for which other reciprocal lattice points are greater than a given distance from the Ewald sphere.

It is well known (see p. 32) that systematic double and triple diffraction occur for the normal-beam and equi-inclination cases respectively, if the crystal is oriented with a symmetry axis parallel to the goniometer-head axis (ϕ-axis). According to Burbank (1965), the situation is even worse for the single crystal orienter technique (four circle diffractometer), although the systematic diffraction he describes will occur only in the special symmetrical-A setting and is avoided in the general setting.

We conclude that the occurrence of simultaneous reflexions is minimized by carefully selecting the azimuthal angle ψ of the primary reflexion, although it may not be possible to avoid simultaneous reflexions altogether. On the other hand, if the exchange of energy between the reflexions is small and measurements of the highest accuracy are not required, their occurrence may be deliberately exploited to speed up the collection of intensity data by using multiple-counter methods (see p. 56).

CHAPTER 10

PROCEDURE FOR MEASURING INTEGRATED INTENSITIES

In this chapter we shall describe the main features in the procedure for measuring the set of relative integrated intensities ρ_{hkl} of the crystal. We shall assume that the diffractometer has been correctly positioned and alined with respect to the incident beam. The methods of making these adjustments are fully described by Furnas (1957).

10.1. Choice of crystal size and shape

To minimize errors arising from the effects of absorption, extinction and simultaneous reflexions, the crystal must be as small as possible. The ultimate limit to the specimen size is determined by counting statistics (see §10.7), and, because of the greater intensity of primary X-ray beams compared with monochromatic neutron beams, the crystal size tends to be much larger in neutron diffraction. In the early period of neutron diffraction the linear dimensions of the crystal were measured in centimetres, but now, with the availability of high-flux reactors and with the use of automatic methods to speed up the collection of data, the size of crystal is down to 1 or 2 mm.

On p. 240 we gave the criterion $\mu R \sim 1$ as a rough indication of the optimum size of the crystal, where μ is the linear absorption coefficient and R the average crystal radius. From Table XIX, this corresponds to an average radius for X-ray work of about 0·1 mm. The diffracted intensity decreases with the size of the unit cell, so that larger specimens are used when the cell size is large and the absorption is relatively low.

The ideal shape of crystal is a sphere for three-dimensional diffraction measurements and a cylinder for two dimensions. With such a crystal it is not only possible to calculate the correction due to absorption, but also to minimize errors arising from non-uniformity of the incident beam. A widely used method for cutting spherical crystals is that devised by Bond (1951) and later improved by Belson (1964). The crystal is placed in a circular tunnel, whose inside periphery is lined with fine abrasive material, and is tumbled randomly against the walls of the tunnel by a jet or jets of compressed air. A cylindrical crystal can be prepared by directing a fine jet of abrasive powder against the crystal as it turns about its axis.

10.2. Alinement of crystal

By 'crystal alinement' we mean the centring of the crystal and the setting of a principal crystallographic axis along the ϕ-axis of the diffractometer. (It is always necessary to centre the crystal, but with a four-circle diffractometer measurements can proceed without further adjustments to the crystal, p. 51.) The crystal is 'centred', if it remains in the same portion of the incident beam while being turned from one Bragg reflexion to the next. The alinement is carried out conveniently by mounting the crystal on a goniometer head which has two arcs, two translational movements and a height adjustment: suitable designs of goniometer head are described on p. 79.

Let us suppose that the crystal is monoclinic and that it is to be alined with the unique axis (diad axis) parallel to the goniometer-head axis, or the ϕ-axis, of a four-circle diffractometer. If the diad axis is along [010], then the normal to the {0k0} planes lies along the ϕ-axis. At this orientation the intensity of the 0k0 reflexions will be unchanged as the crystal turns through 360° about the ϕ-axis. This is not quite true if systematic errors are present of the type discussed in the previous chapter, but for correct alinement there is a minimum variation of the diffracted intensity with the angle ϕ, and the intensity will not fall to zero at any point in the complete rotation about ϕ.

The alinement proceeds, therefore, as follows. The crystal is viewed through a telescope and is centred using the translational movements of the goniometer head. The ϕ-axis of the four-circle diffractometer is then moved into the horizontal plane containing the incident beam by rotating it around the χ-circle (see Fig. 113a). The goniometer head is turned about the ϕ-axis to bring one of the arcs into the horizontal plane and that arc is adjusted for maximum

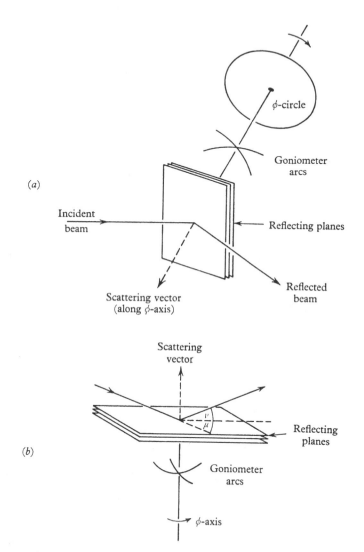

Fig. 113. (a) Alinement of crystal on four-circle diffractometer: incident and reflected beams are in horizontal plane. (b) Alinement of crystal on equi-inclination diffractometer: incident and reflected beams are in vertical plane.

intensity of the $ok0$ reflexion. The head is rotated through 180° and the same arc is adjusted again for maximum intensity. If the arc settings differ for the two orientations of the goniometer head, the correct setting is taken as the mean. The second arc is adjusted in the same way at the 90° and 270° positions of the goniometer head. For a eucentric goniometer head (see p. 79), with the crystal accurately positioned at the centre of the arcs, the centring will not be affected by any movement of the arcs, but for other mountings alternate movements of cross-slides and arcs are necessary during the alinement. With an equi-inclination instrument a similar alinement procedure is possible (see Fig. 113 b).

A great convenience during the alinement process is the provision of two pairs of masks which can obscure the right or left halves or the top or bottom halves of the aperture of the detector. These masks allow one to judge easily in which direction the arcs need to be moved for correct alinement since they define where the reflexion is, relative to the centre of the detector aperture. The reflexion is correctly centred when the intensities in the four quadrants are approximately equal.

Alinement is much more difficult if the crystal cannot first be approximately oriented and if nothing is known about its unit-cell dimensions. The necessary steps in this situation have been described so fully and clearly by Furnas (1957) that no further comments are needed here.

Wooster (1965) describes a procedure for determining the setting angles of a crystal for a four-circle diffractometer assuming no prior knowledge of its orientation or lattice parameters.

10.3. Correct assignment of signs of crystallographic axes

For a crystal of any symmetry it is not possible to distinguish the configuration of the axes **a, b** and **c** from $-\mathbf{a}$, $-\mathbf{b}$ and $-\mathbf{c}$ using the observed positions only of the Bragg reflexions. This distinction is important in determining the absolute configuration of a non-centrosymmetrical crystal, and can be made from the intensities of the hkl, $\bar{h}\bar{k}\bar{l}$ pairs of reflexions (Ramaseshan, 1964). Let us suppose, for instance, that the intensities of 200 and $\bar{2}$00 are related by the inequality $I_{200} > I_{\bar{2}00}$. If the crystal is replaced by a different crystal of the same material, the second crystal must be indexed in such a way that the same inequality holds; should the initial assignment of axes lead to $I_{200} < I_{\bar{2}00}$, it will be necessary to invert the **a**-, **b**- and **c**-axes.

We referred on p. 80 to the problem of determining the correct sense, $+\mathbf{b}$ or $-\mathbf{b}$, of the diad axis in a monoclinic crystal relative to the vectors **a**, **c**, when the crystal is mounted with the **b**-axis

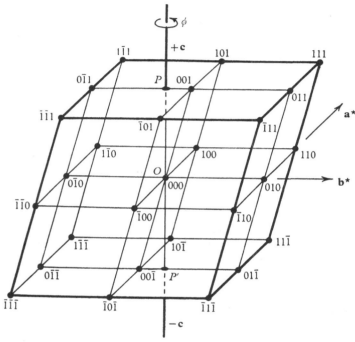

Fig. 114. Monoclinic reciprocal lattice with unique axis along **b** or **b*** and rotation axis along **c**. If the sense of **c** is chosen incorrectly the point 001 in the $l = 1$ level will lie on the other side of the ϕ axis, OP.

parallel to the goniometer-head axis ϕ. If the **a**- or **c**-axes of the crystal are along the ϕ-axis, the correct sense of **b** cannot be found from observations in the zero level alone (see Fig. 114). However, the observed positions of reflexions in the higher level will immediately reveal any wrong choice of sign, which can be corrected by inverting the crystal or reversing the sense of rotation of the ϕ-shaft: this may be more convenient than recomputing the setting angles with a revised assignment of signs.

10.4. Measurement of lattice parameters

It is necessary to determine the lattice parameters of the crystal before calculating the setting angles for each reflexion, using the formulae given in Chapter 2.

The lattice parameters can be measured with a diffractometer to a precision equal to that with the best X-ray camera, and the same instrument can then be used to measure the integrated intensities without further adjustment of the crystal. The crystal is alined on the diffractometer (for instance, with the goniometer-head axis lying along the normal to a family of reflecting planes), and the lattice parameters are derived from the observed scattering angles $2\theta_{obs.}$ for a number of different reflexions.

There are at least two ways of measuring 2θ with a four-circle diffractometer. In the first (Fig. 115 a) the reflexion is observed with a wide detector aperture and the profile of the reflexion is plotted using an ω-scan. The crystal is then set at the peak position, the detector aperture is closed down to a narrow vertical slit, and a second intensity scan is performed with the crystal stationary and the detector moving. $2\theta_{obs.}$ is taken as the difference between two readings of the angular position of the detector, one at the peak of the reflexion and the other at the point of maximum intensity of the direct beam, suitably attenuated.

In the second method of determining 2θ, there is no need to measure the 'straight-through' position of the direct beam. The *hkl* reflexion is observed with the ω shaft of the crystal and the θ shaft of the detector coupled together in a 1:2 angular ratio. The detector aperture remains wide open and so the reflexion need not enter the detector centrally. The value of ω is noted for the peak position of the reflexion. With ω and θ remaining coupled, the ω-axis is then turned so that the beam enters the detector by reflexion at the reverse side of the *hkl* reflecting plane (see Fig. 115). $2\theta_{obs.}$ is the difference in the readings of the ω scale for the peak positions of *hkl* and *h̄k̄l̄*. By measuring the angle between the two positions of the crystal, rather than the angle between the two detector positions as in the first method, not only the zero error but also errors due to absorption and incorrect centring of the crystal are eliminated: absorption and eccentricity

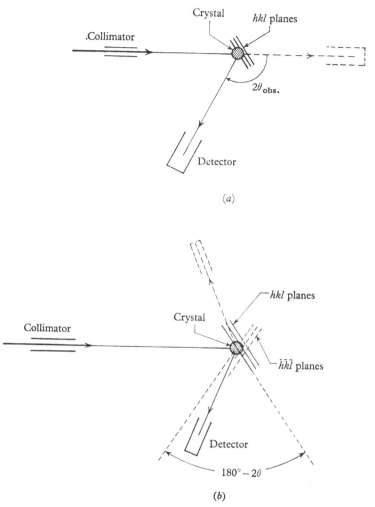

Fig. 115. Two ways of measuring scattering angle 2θ. The zero position of the detector arm ($2\theta = 0$) must be known in (a), but not in (b).

both alter the angle at which the beam enters the detector, but the crystal angles are still correct if the detector window is wide open. Bond (1960) gives further details of this second method and discusses the various sources of systematic error. He shows that the scattering angles of a perfect crystal, such as silicon, can be measured with X-rays by this method to better than one part in 10^5. Further development of this method by Baker (1966) gives results accurate to one part in 10^7.

After the measurement of the scattering angles for a number of reflexions, the lattice parameters of the reciprocal unit cell can readily be determined by a least-squares analysis, using equation (2.3) in the form:

$$\frac{4\sin^2\theta}{\lambda^2} = h^2a^{*2} + k^2b^{*2} + l^2c^{*2} + 2hka^*b^*\cos\gamma^*$$
$$+ 2klb^*c^*\cos\alpha^* + 2lhc^*a^*\cos\beta^*. \quad (10.1)$$

The observed quantities are the set of 2θ values and the variables are the six lattice parameters. The final accuracy of these parameters is related to the precision with which the wavelength λ is known: this precision is lower in neutron diffraction than in X-ray work because of the larger bandwidth $\delta\lambda$ of the incident neutron beam.

Table XXIII illustrates some results obtained by Busing (1965), who analysed X-ray diffractometer data from a single crystal of monoclinic barium chloride dihydrate, $BaCl_2 . 2H_2O$. Twelve independent values of 2θ were determined by the method illustrated in Fig. 115a, and these observed values were compared with the calculated values assuming a triclinic unit cell. The values of α and γ refined to within one standard deviation of the expected monoclinic values of 90°.

TABLE XXIII. *Unit cell dimensions of $BaCl_2 . 2H_2O$ from least-squares refinement of X-ray data (after Busing, 1965)*

Initial values of lattice parameters before least-squares refinement	Final values of lattice parameters
$a = 6{\cdot}7380$ Å	$6{\cdot}7215 \pm 0{\cdot}0002$ Å
$b = 10{\cdot}8600$ Å	$10{\cdot}9077 \pm 0{\cdot}0003$ Å
$c = 7{\cdot}1360$ Å	$7{\cdot}1315 \pm 0{\cdot}0003$ Å
$\cos\alpha = 0$	$-0{\cdot}00004 \pm 0{\cdot}00007$
$\cos\beta = -0{\cdot}0166$	$-0{\cdot}01924 \pm 0{\cdot}00006$
$\cos\gamma = 0$	$-0{\cdot}00007 \pm 0{\cdot}00006$

10.5. Choice of scan

On p. 20 we described the three measuring procedures which may be used to determine the intensity of a Bragg reflexion. These are the stationary-crystal-stationary-detector method, the moving-crystal-stationary-detector method (ω-scan), and the moving-crystal-moving-detector method ($\omega/2\theta$-scan).

The simplest procedure is the stationary-crystal-stationary-detector method which requires a uniform convergent incident beam (see Fig. 7, p. 22). Unfortunately, this requirement is difficult to meet in both X-ray and neutron diffraction.

Alexander & Smith (1962) have examined the relation between peak height and peak area when the incident beam is not sufficiently convergent to give flat-topped peaks. They showed that the peak height is proportional to the peak area in a very limited range of θ only. The same relation has been examined for neutron diffraction by Chidambaram, Sequeira & Sikka (1964) who claimed that the data collection rate can be speeded up by a factor of about 10 using peak heights and an experimental calibration curve relating peak height and peak area.

For accurate intensity measurements stationary-crystal methods are not recommended and in most investigations the choice of measuring procedures will be between the ω-scan and the $\omega/2\theta$-scan.

As Burbank (1964) has pointed out, the preferred scan is that which yields the true integrated intensity by illuminating the smallest volume in reciprocal space: the background is then as low as possible, especially that part which is contributed by thermal diffuse scattering and which tails off slowly at the sides of the Bragg reflexion (p. 221). Using this criterion of smallest volume in reciprocal space, the $\omega/2\theta$-scan is generally preferred in neutron work, because of the large bandwidth $\delta\lambda$ of the incident radiation (see Fig. 116a). On the other hand, if the mosaic spread of the crystal is large, the ω-scan may be preferable at low Bragg angles (see Fig. 116b), where the spectral dispersion is least.

Both the horizontal and vertical dimensions of the detector aperture must be carefully chosen for both scans. These dimensions vary with the angle θ and, in the case of inclination diffractometers, with the level of the reciprocal lattice under investigation. They are best found experimentally by examining the integrated intensities of a number of reflexions as a function of the aperture size, and selecting the minimum size at a particular θ which gives the full integrated intensity.

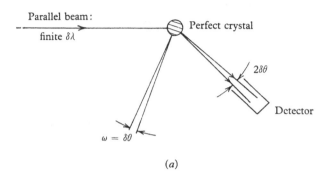

(a)

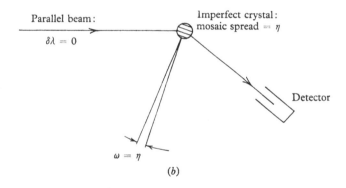

(b)

Fig. 116. Diagrams illustrating use of: *(a)* $\omega/2\theta$-scan (2θ-scan) for perfect crystal illuminated by radiation with spectral range $\delta\theta = \tan\theta\,\delta\lambda/\lambda$; *(b)* ω-scan for crystal with large mosaic spread η.

Alexander & Smith (1962, 1964) and Burbank (1964) have derived theoretical expressions for the minimum aperture dimensions in terms of the mosaic spread of the crystal and the divergence and spectral dispersion of the incident beam. Burbank's results are summarized in Table XXIV using the notation adopted in Chapter 6. The table includes the minimum value of the rocking range of the crystal which was discussed on p. 174.

TABLE XXIV. *Comparison of ω and $\omega/2\theta$ scans*

	ω-Scan	$\omega/2\theta$-Scan
Minimum horizontal detector aperture	$2\delta\theta + \delta_F^H + 2\delta_C^H \cos^2\theta$	$\delta_F^H + 2\delta_C^H \sin^2\theta + \eta$
Minimum vertical detector aperture	$\delta_F^V + \delta_C^V + \eta$	$\delta_F^V + \delta_C^V + \eta$
Minimum rocking range	$\delta_F^H + \delta_C^H + \eta + \delta\theta$	$\delta_F^H + \delta_C^H + \eta + \delta\theta$
Preferred scan using		
(a) X-rays with balanced filters	—	Always preferred
(b) X-rays or neutrons with monochromator, X-rays with simple filtration	At small Bragg angles when η is large	Always at large θ; at small θ when η is small

The quantities $\delta\theta$, δ_F, δ_C and η are defined on p. 174. The superscripts H and V refer to the horizontal and vertical planes, respectively.

10.6. Strategy for exploring reciprocal space

With an equi-inclination instrument it is convenient to measure the reflexions layer-by-layer in reciprocal space. It is then necessary to alter only one angular setting (ϕ) for the crystal and one (Υ) for the detector in moving from one reflexion to the next within the same layer. This procedure also involves minimum alteration of the size of the detector aperture, which must be changed in moving to the upper layers.

With a four-circle diffractometer the strategy adopted in exploring reciprocal space will depend on the setting speeds of the diffractometer shafts. When the time taken for shaft setting is an appreciable fraction of the total measuring time, the reflexions must be measured in a sequence which minimizes the angular increments between one reflexion and the next; in practice, a zigzag scan along neighbouring reciprocal lattice lines does this quite effectively. When the setting speed is high enough for the setting time to become small compared with the measuring time (as is nearly always the case in neutron diffraction), it becomes better to measure the reflexions in the order of increasing Bragg angle. This has the advantage that the angular range of scan for each reflexion and the correct size of the detector aperture, both of which are dependent on θ only, can be kept fixed for all reflexions within a limited range of θ. Furthermore, the maximum observed Bragg angle $\theta_{\text{max.}}$ (or, more accurately, the quantity $\sin\theta_{\text{max.}}/\lambda$) determines the limit of resolution of the atoms in the Fourier map of the unit cell; by recording the reflexions in increasing θ the investigation can be terminated when the desired resolution is reached.

It is advisable to return at regular intervals during the intensity measurements to one or more standard reflexions. These standards are extremely useful in checking the correct functioning of the apparatus (for example, the stability of the detector and of the counting circuits) and in ensuring that the crystal retains its correct alinement during the experiment and that it does not deteriorate in the incident beam. The extra time spent in measuring the standard reflexions is a small price to pay for the satisfaction of knowing that the measurement of the remaining reflexions is proceeding satisfactorily.

10.7. Counting statistics and period of counting

The emission of photons by an X-ray tube, or of neutrons by a nuclear reactor, is a random process and gives rise to a statistical uncertainty in the measurement of the diffracted intensities. In theory, this statistical uncertainty can be reduced to any required level by prolonging the counting process for a sufficient period of time. In practice, however, the total time available for measuring a set of integrated intensities is limited, and it is important to know how best to divide the time so as to obtain the greatest overall precision in the measurements.

We note here the distinction that is drawn between the terms *uncertainty* and *error*. The term *uncertainty* is used to indicate the statistical fluctuation between different measurements of the same quantity made under identical experimental conditions, and is independent of any systematic error present. *Error* refers to the difference between the measured value and the 'real' value of the physical quantity, and includes the effects of both statistical uncertainty and systematic error. The same distinction is made between the terms *precision* and *accuracy*: *precision* indicates the closeness with which measurements agree with one another, whereas *accuracy* denotes the closeness of the measurements to the real value.

Let us assume that the number of counts N, measured in equal times t, fluctuates according to a Gaussian distribution about the mean $\bar{N}$. The standard deviation of the distribution is

$$\sigma(N) = \bar{N}^{\frac{1}{2}},$$

so that each individual determination of N has a relative statistical uncertainty given by

$$\epsilon = Q\sigma/\bar{N} = Q\bar{N}^{-\frac{1}{2}}, \qquad (10.2)$$

where Q is a constant determined by the 'confidence level'. For the 50 per cent confidence level, representing a 50 per cent probability that $N - \bar{N}$ is less than ϵ, Q is 0·67. Q is 1·64 for the 90 per cent level and 2·58 for the 99 per cent level. Fig. 117 shows the percentage uncertainty as a function of N for the 50, 90 and 99 per cent confidence levels. For a percentage uncertainty of 1 per cent the total number of counts to be accumulated is 4,500 for the 50 per cent level, 27,000 for the 90 per cent level, and 67,000 for the 99 per cent level. Thus, to raise the confidence level from 50 to 99 per cent the total number of accumulated counts must be multiplied fifteen times.

Effect of background and optimization of counting period

The magnitude of the integrated intensity is determined as the difference between two counts, N_1 and N_2, where N_1 is the count for the Bragg peak and N_2 is the background count. Assuming equal times on peak and background the integrated intensity is ρ, where

$$\rho \propto N_1 - N_2.$$

The standard deviation of the difference is given by

$$\sigma = (\sigma_1^2 + \sigma_2^2)^{\frac{1}{2}}, \qquad (10.3)$$

where σ_1 and σ_2 are the standard deviations of the two individual counts. Thus, if in a certain experiment the overall background count is 100 and that due to the peak-plus-background is 200, the standard deviation of the difference is

$$\sigma = (200 + 100)^{\frac{1}{2}} = 17.$$

For a zero background, the peak-plus-background would be recorded as 100, and the standard deviation would be only 10. The percentage standard deviation, $100\sigma/(N_1 - N_2)$, is given in Table XXV for different values of N_1 and N_2.

TABLE XXV. Percentage standard deviation of $(N_1 - N_2)$ as a function of N_1 and N_2

Values of N_1

N_2	1,000	2,000	3,000	4,000	5,000	6,000	7,000	8,000	9,000	10,000	12,000	15,000	20,000
100	3·7	2·4	1·9	1·6	1·5	1·3	1·2	1·1	1·1	1·0	0·9	0·8	0·7
200	4·3	2·6	2·0	1·7	1·5	1·4	1·2	1·1	1·1	1·0	0·9	0·8	0·7
300	5·1	2·8	2·1	1·8	1·5	1·4	1·3	1·1	1·1	1·1	0·9	0·8	0·7
400	6·2	3·1	2·2	1·8	1·6	1·4	1·3	1·2	1·1	1·1	1·0	0·9	0·7
500	7·7	3·3	2·4	1·9	1·7	1·5	1·3	1·2	1·2	1·1	1·0	0·9	0·7
600	10·0	3·7	2·5	2·0	1·7	1·5	1·3	1·3	1·2	1·1	1·0	0·9	0·7
700	13·7	4·0	2·6	2·1	1·8	1·5	1·4	1·3	1·2	1·1	1·0	0·9	0·7
800	21·2	4·4	2·8	2·2	1·8	1·6	1·4	1·3	1·2	1·2	1·0	0·9	0·8
900	43·5	4·9	3·0	2·3	1·9	1·6	1·5	1·3	1·2	1·2	1·0	0·9	0·8
1,000	—	5·5	3·2	2·4	2·0	1·7	1·5	1·4	1·3	1·2	1·0	0·9	0·8
1,100	—	6·2	3·3	2·5	2·0	1·7	1·5	1·4	1·3	1·2	1·1	0·9	0·8
1,200	—	7·1	3·6	2·6	2·1	1·8	1·6	1·4	1·3	1·2	1·1	0·9	0·8
1,300	—	8·2	3·9	2·7	2·1	1·8	1·6	1·4	1·3	1·2	1·1	0·9	0·8
1,400	—	9·7	4·1	2·8	2·2	1·9	1·6	1·5	1·3	1·2	1·1	0·9	0·8
1,500	—	11·8	4·5	3·0	2·3	1·9	1·7	1·5	1·4	1·3	1·1	1·0	0·8
1,600	—	15·0	4·9	3·1	2·4	2·0	1·7	1·5	1·4	1·3	1·1	1·0	0·8
1,700	—	20·3	5·2	3·3	2·5	2·0	1·7	1·6	1·4	1·3	1·1	1·0	0·8
1,800	—	30·8	5·8	3·5	2·6	2·1	1·8	1·6	1·4	1·3	1·2	1·0	0·8
1,900	—	62·5	6·4	3·7	2·7	2·2	1·8	1·6	1·5	1·4	1·2	1·0	0·8
2,000	—	—	7·1	3·9	2·8	2·2	1·9	1·7	1·5	1·4	1·2	1·0	0·8
3,000	—	—	—	8·4	4·5	3·1	2·5	2·1	1·8	1·6	1·4	1·1	0·9
4,000	—	—	—	—	9·5	5·0	3·5	2·7	2·3	2·0	1·6	1·3	1·0
5,000	—	—	—	—	—	10·5	5·5	3·8	3·0	2·5	1·9	1·5	1·1

We must now consider the problem: given counting rates n_1, n_2 which are determined by counting for periods T_1, T_2 respectively on a reflexion, how should the total time $T = T_1 + T_2$ be divided so that the

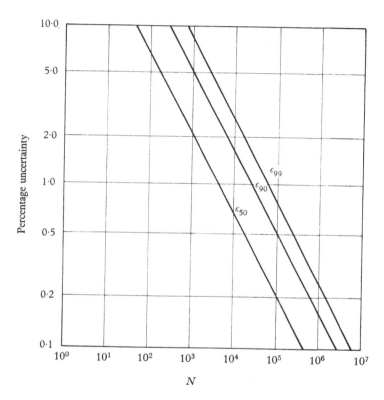

Fig. 117. Percentage statistical uncertainty as a function of the total number of counts N. The straight lines refer to the 50, 90, 99 per cent confidence levels.

percentage statistical uncertainty in ρ $(=n_1-n_2)$ is a minimum? The following treatment of this problem is similar to that of Mack & Spielberg (1958).

Let ϵ_1 and ϵ_2 be the relative statistical uncertainties in n_1 and n_2. The absolute magnitudes of these uncertainties are $n_1\epsilon_1$ and $n_2\epsilon_2$ and the uncertainty of the difference $n_1 - n_2$ is

$$[(n_1\epsilon_1)^2 + (n_2\epsilon_2)^2]^{\frac{1}{2}}.$$

The relative uncertainty of ρ, therefore, is

$$\epsilon = \frac{[(n_1\epsilon_1)^2 + (n_2\epsilon_2)^2]^{\frac{1}{2}}}{n_1 - n_2}. \tag{10.4}$$

The total number of counts accumulated in measuring the peak-plus-background is n_1T_1 and the number for the background alone is n_2T_2. Thus, from equation (10.2),

$$\epsilon_1 = Q(n_1T_1)^{-\frac{1}{2}},$$

and $$\epsilon_2 = Q(n_2T_2)^{-\frac{1}{2}}.$$

Substituting in equation (10.4) gives

$$\epsilon = \frac{Q}{n_1 - n_2}\left(\frac{n_1}{T_1} + \frac{n_2}{T_2}\right)^{\frac{1}{2}}. \tag{10.5}$$

n_1, n_2 are fixed counting rates and ϵ is a function of the two variables T_1, T_2:

$$\epsilon = \epsilon(T_1, T_2). \tag{10.6}$$

Our problem is now reduced to finding the minimum value of ϵ, subject to the restraining condition

$$T = T_1 + T_2. \tag{10.7}$$

From (10.6), the minimum condition is

$$\frac{\partial \epsilon}{\partial T_1}dT_1 + \frac{\partial \epsilon}{\partial T_2}dT_2 = 0, \tag{10.8}$$

and from (10.7) the condition $T = $ constant gives

$$dT_1 + dT_2 = 0. \tag{10.9}$$

Combining (10.8) and (10.9):

$$\frac{\partial \epsilon}{\partial T_1} = \frac{\partial \epsilon}{\partial T_2}, \tag{10.10}$$

and substituting equation (10.5) for ϵ into (10.10) yields

$$\left(\frac{T_1}{T_2}\right)^2 = \frac{n_1}{n_2} = K, \tag{10.11}$$

where K is the ratio of the counting rates. Finally, (10.7) and (10.11) give

$$T_1 = \frac{K^{\frac{1}{2}}}{1+K^{\frac{1}{2}}}\,T,$$
$$T_2 = \frac{1}{1+K^{\frac{1}{2}}}\,T. \qquad (10.12)$$

Equations (10.12) describe the optimum division of time T between T_1 and T_2. For a weak reflexion, with the peak counting rate only slightly higher than the background rate, $K \approx 1$ and equal times should be spent on the peak and on the background. On the other hand, if the peak-to-background ratio is very high, only a small proportion of the time should be spent in measuring the background.

We ask next: given an optimum division of the total time T between the peak and the background, what is the minimum value of T to achieve a stipulated level of precision in the integrated intensity ρ? The answer to this second question determines the optimum division of time between all the reflexions which are to be recorded in a particular investigation.

By manipulating equations (10.5) and (10.11) we arrive at

$$N_1 = n_1 T_1 = \frac{Q^2}{\epsilon^2}K^{\frac{3}{2}}\frac{K^{\frac{1}{2}}+1}{(K-1)^2},$$

and

$$N_2 = n_2 T_2 = \frac{Q^2}{\epsilon^2}\frac{K^{\frac{1}{2}}+1}{(K-1)^2}. \qquad (10.13)$$

These equations give the total number of counts accumulated on the peak, N_1, and on the background, N_2, as a function of the relative uncertainty ϵ in measuring $n_1 - n_2$. Note that N_1 and N_2 are each independent of the total time T spent in measuring the reflexion: to attain a specific precision, corresponding to a fixed value of Q/ϵ, a predetermined number of counts must be accumulated, irrespective of the counting time.

Table XXVI gives values of N_1, N_2, T_1, T_2, for a fixed background counting rate n_2 of 10 counts/s and for different values of the peak-to-background ratio n_1/n_2. These quantities are calculated from equations (10.13) for a probable uncertainty of 1 per cent in the integrated intensity. (The probable uncertainty corresponds to a confidence level of 50 per cent, or $Q = 0\cdot67$.)

18

A & W

TABLE XXVI. *Total time for measuring $n_1 - n_2$ to a probable relative uncertainty of 1 per cent ($n_2 = 10$ counts/s)*

n_1 (counts/s)	$\frac{n_1}{n_2} = K$	N_1 (counts)	N_2 (counts)	T_1 (s)	T_2 (s)	$T = T_1 + T_2$
12	1·2	310,000	240,000	26,000	24,000	14 h.
15	1·5	73,000	40,000	4,900	4,000	2½ h.
20	2	31,000	11,000	1,600	1,100	45 min.
40	4	12,000	1,500	300	150	7½ min.
100	10	6,000	190	60	19	79 s.
1,000	100	5,000	5	5	0·5	5·5 s.

The table shows that the total time for measuring a weak reflexion with a peak-to-background ratio of 2:1 is 34 times longer than the corresponding time for a strong reflexion with a ratio of 10:1. Obviously, a very considerable economy of time results, provided all reflexions are to be measured to about the same relative precision, by reducing the time spent in measuring the strong reflexions. It is also worth emphasizing again the crucial importance of reducing the background level as much as possible. The background level, and not the absolute magnitude of the Bragg peak above this level, sets the main limit to the magnitude of the reflexion which can be observed.

Statistics of monitoring

When a monitoring system is employed to register the strength of the incident radiation, statistical fluctuations in the monitored beam affect the measured intensity of the diffracted beam. If N is the number of diffracted pulses from the specimen, and these are recorded in the time taken to count N_0 pulses in the monitoring counter, the standard deviation of the measured counting rate from the specimen is

$$N^{\frac{1}{2}}(1 + N/N_0).$$

Thus, provided N_0 is at least ten times greater than N, the effect of the monitoring statistics can be ignored.

Time optimization by on-line computer control

We have discussed the optimization of measuring time between the different reflexions of the crystal, and between the peak and background of an individual reflexion, using the criterion that the minimum total time should be spent in measuring all the integrated intensities to a given *relative* precision.

A different division of time occurs if we measure the structure factors F to a constant *absolute* precision: thus *if we were able to neglect the background* we should have to spend equal times on each reflexion. This is shown by writing the integrated intensity ρ as

$$\rho = n = cF^2, \tag{10.14}$$

where n is the peak counting rate and c is a constant incorporating the Lorentz-polarization conversion factor (see Chapter 11). If T is the time of counting for one reflexion, the total number of counts accumulated is nT and the uncertainty in this number is $(nT)^{\frac{1}{2}}$. Thus the uncertainty in the counting rate is $(nT)^{\frac{1}{2}} \div T$, that is

$$\sigma(\rho) = \sqrt{\frac{n}{T}}.$$

From equation (10.14)

$$\sigma(\rho) = 2cF\sigma(F),$$

or

$$\sigma(F) = \frac{\sigma(\rho)}{2cF} = \frac{1}{2c} \cdot \sqrt{\frac{n}{T}} \sqrt{\frac{c}{n}}$$

$$= \frac{1}{2}\left(\frac{1}{cT}\right)^{\frac{1}{2}}.$$

$\sigma(F)$, therefore, is independent of n: its magnitude is determined by the period of counting T and is the same for all values of F. This contrasts with the corresponding condition for $\sigma(F)/F$ to be constant, which is that counting continues until a fixed number of counts nT is accumulated.

To determine the optimum division of time between the different reflexions we require to know the values of F, the very quantities we aim to measure. One way around this difficulty is to make a rapid preliminary set of measurements to a low statistical accuracy to determine the counting strategy for a final run. A more sophisticated approach is possible when the diffractometer is connected on-line to a computer.

Let us suppose that the total time for measuring a reflexion is, on average, ten minutes and that the integrated intensity is to be measured to a relative uncertainty $\sigma(\rho)/\rho$ $[= \sigma(F)/F]$ of 5 per cent. By scanning quickly across the reflexion in one minute and registering N_1 counts on the peak and N_2 counts on the background, the quantity

$$\frac{\sigma(\rho)}{\rho} = \frac{(N_1 + N_2)^{\frac{1}{2}}}{(N_1 - N_2)}$$

can be computed at the end of the scan. The scan is then repeated, the total counts added to those found in the first scan, and $\sigma(\rho)/\rho$ recomputed. The repeated scanning continues until the value of $\sigma(\rho)/\rho$ falls to 5 per cent, and the control equipment then switches to the next reflexion. Table XXVII indicates the flow-sheet for measuring the reflexions in this way.

TABLE XXVII. *Flow-sheet for measuring reflexions to constant*
$$\sigma(\rho)/\rho = 5\,\%$$

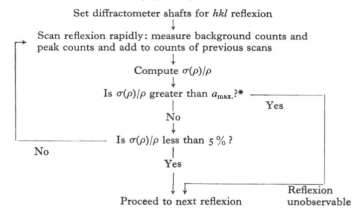

* The value of $a_{max.}$ is, perhaps, 100 %: its precise value is dependent on the maximum time spent on measuring the weakest reflexions.

CHAPTER II

DERIVATION AND ACCURACY OF STRUCTURE FACTORS

In the previous chapter we described the procedure for measuring the relative integrated intensities of a single crystal. If an infinitesimally small block of volume δV reflects the incident X-ray beam, it can be shown (see, for example, Buerger, 1960) that the absolute magnitude of the integrated intensity is proportional to δV:

$$\rho_{hkl} = Q \, \delta V,$$

where the constant of proportionality Q is

$$Q = \left(\frac{e^2}{mc^2}\right)^2 N_c^2 \lambda^3 Lp \, |F_{hkl}|^2. \qquad (11.1)$$
$$\text{(X-rays)}$$

In this expression N_c is the number of unit cells per unit volume, λ the wavelength, and F_{hkl} the structure factor: the Lorentz factor L and polarization factor p are angle factors, which depend on the experimental arrangement used in measuring the intensities, and whose forms are discussed in §§11.1 and 11.2. In general, the structure factor F_{hkl}, characterizing the wave scattered by the (hkl) plane, is a complex quantity: its magnitude $|F_{hkl}|$ only is related to the observed intensity. The determination of its phase, which can have any value between $0°$ and $360°$ relative to a wave scattered at the origin of the unit cell, constitutes the familiar 'phase problem'.

In neutron diffraction the nuclear scattering amplitude is analogous to the X-ray atomic scattering factor multiplied by e^2/mc^2 (Bacon, 1962). Consequently, equation (11.1) applies also to neutrons with $(e^2/mc^2)^2$ replaced by unity:

$$Q = N_c^2 \lambda^3 Lp \, |F_{hkl}|^2. \qquad (11.2)$$
$$\text{(neutrons)}$$

The experimental values of ρ_{hkl} are measured on a relative scale: to measure their absolute magnitudes the appropriate scale factor is determined in a separate experiment, as described in §11.3.

Thus we can express both (11.1) and (11.2) more conveniently in the form

$$\rho_{hkl} = cLp|F_{hkl}^{\text{obs.}}|^2, \qquad (11.3)$$

where c is a single scale factor for all the reflexions and the superscript 'obs.' indicates observed structure factors.

Later in this chapter we shall compare the observed and calculated structure factors for crystals of known simple structure, quoting results obtained with X-rays and with neutrons. This comparison gives some indication of the accuracy to be expected in measuring with a diffractometer the structure factors of any crystal. (It is worth emphasizing that in the examination of complex structures, such as protein crystals, speed of data collection is more important than extreme accuracy: many thousands of reflexions are measured, often while the crystal is deteriorating in the X-ray beam, and the final accuracy of the structure factors is determined primarily by counting statistics (see p. 268) and not by systematic errors.) First, however, we shall describe the evaluation of the three terms L, p, c, in equation (11.3), which allow the conversion of the integrated intensities to a set of observed structure factors.

11.1. Lorentz factor

The intensity of a reflexion is proportional to the time during which the corresponding reciprocal lattice point is close to the surface of the reflecting sphere: the Lorentz factor is a geometrical term which corrects for the different rates at which the reciprocal lattice points sweep through the sphere. If ω is the angular velocity of the crystal and v_n is the component of the velocity of the reciprocal lattice point along the radius of the sphere, the Lorentz factor is defined as the ratio of these velocities:

$$L = \omega/v_n. \qquad (11.4)$$

The expression for L in terms of measurable crystal and instrumental angles depends on the particular diffraction geometry used.

Normal-beam equatorial geometry (see §2.4)

Both the incident and reflected beams are normal to the vertical axis of rotation through O (Fig. 118). The linear velocity of the reciprocal lattice point $P(= hkl)$ is $\omega\xi$ along a direction normal to the radial co-ordinate ξ, and the velocity normal to the sphere of reflexion is

$$v_n = \omega\xi\cos\theta.$$

The radius of the circle of reflexion is unity, so that

$$\xi = 2\sin\theta$$

and the inverse Lorentz factor is

$$L^{-1} = 2\sin\theta\cos\theta = \sin 2\theta. \qquad (11.5)$$

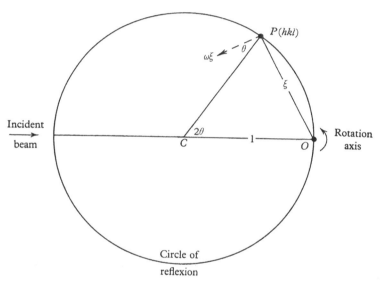

Fig. 118. Derivation of Lorentz factor for normal-beam equatorial geometry.

The variation of L with θ is shown in Fig. 119. At low Bragg angles, L is large because the reciprocal lattice point is close to the origin of the reciprocal lattice; and at high θ, L is large because the *hkl* point passes tangentially, or nearly so, through the surface of the reflecting sphere. The Lorentz factor reduces the intensities by the maximum amount at $\theta = 45°$.

General inclination geometry (see §2.3)

The Lorentz factor assumes a very simple form for the equatorial method, but this is not so for inclination geometry, where the incident beam is inclined at a variable angle $90° - \mu$ to the axis of rotation and the reflected beam at a variable angle $90° - \nu$.

To derive an expression for L we refer to Fig. 120a. If ω is the angular velocity of the crystal about the axis OO_l, the reciprocal lattice point P in the l-level has a linear velocity $\omega\xi$ along the

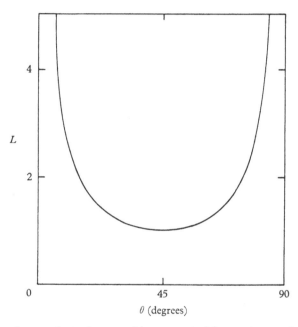

Fig. 119. Lorentz factor for normal-beam equatorial geometry, as a function of θ.

direction PU: PU is a vector in the l-level which is perpendicular to PO_l. The component of this velocity normal to the sphere of reflexion is

$$v_n = \omega\xi\cos r,$$

where r is the angle between PU and the line PC joining the hkl point and the centre of the sphere of reflexion. The inverse Lorentz factor is, therefore,

$$L^{-1} = v_n/\omega = \xi\cos r. \tag{11.6}$$

We obtain an expression for $\cos r$ in equation (11.6) as follows. PU is normal to the radial co-ordinate ξ and lies in the l-level whose circle of reflexion has centre C_l. Let us choose the position

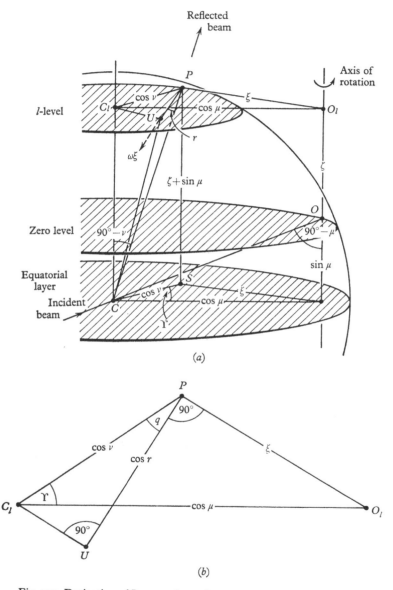

Fig. 120. Derivation of Lorentz factor for general inclination geometry: (a) portion of sphere of reflexion; (b) points in l-level.

of U so that it lies at the foot of the perpendicular from C_l to PU; then PU is horizontal and normal to the vertical plane $C_l CU$ and to every line in that plane. In particular the angle between PU and UC is $90°$ and

$$\cos r = PU/PC = PU. \qquad (11.7)$$

Now from Fig. 120b,

$$PU = PC_l \cos q = \cos \nu \cos q, \qquad (11.8)$$

where q, the angle between PC_l and PU, is given by

$$\frac{\sin(q+90°)}{\cos\mu} = \frac{\sin\Upsilon}{\xi},$$

or $$\cos q = \frac{\cos\mu\sin\Upsilon}{\xi}. \qquad (11.9)$$

Combining (11.7), (11.8), (11.9):

$$\cos r = \frac{\cos\mu\cos\nu\sin\Upsilon}{\xi},$$

and substituting into (11.6) we obtain finally

$$L^{-1} = \cos\mu\cos\nu\sin\Upsilon. \qquad (11.10)$$

The angles μ, ν, Υ are the setting angles of the diffractometer. The expression (11.10) is readily evaluated for each reflexion as part of the computer program for reducing the intensity data to observed structure factors.

Equation (11.5) for the Lorentz factor in the equatorial method can be considered as a special case of the general formula (11.10), with $\mu = \nu = 90°$ and $\Upsilon = 2\theta$. The general formula also reduces to simpler forms for the special settings of inclination geometry.

(a) *Normal-beam setting* ($\mu = 0$). The incident beam is normal to the rotation axis ($\mu = 0$) and (11.10) reduces to

$$L^{-1} = \cos\nu\sin\Upsilon. \qquad (11.11)$$

From equation (2.20)

$$\cos\Upsilon = \frac{2 - \xi^2 - \zeta^2}{2(1-\zeta^2)^{\frac{1}{2}}}, \quad \Big\}$$
and $$\sin\nu = \zeta, \qquad (11.12)$$

where ξ, ζ are the cylindrical co-ordinates of the reciprocal lattice point hkl. Table I on p. 27 lists ξ, ζ in terms of h, k, l, and the lattice parameters of the crystal. L is computed for a crystal of any symmetry using the equations (11.11), (11.12) in conjunction with Table I.

(b) *Equi-inclination setting* ($\mu = -\nu$). The incident and reflected beams are equally inclined to the rotation axis, with $\mu = -\nu$. Thus $\cos\mu = \cos\nu$ and the inverse Lorentz factor, equation (11.10), is

$$L^{-1} = \cos^2\nu \sin\Upsilon. \qquad (11.13)$$

From equation (2.21)

$$\sin\nu = \tfrac{1}{2}\zeta,$$

and

$$\sin\tfrac{1}{2}\Upsilon = \frac{\xi}{2\cos\mu}, \qquad (11.14)$$

and L is computed for the *hkl* reflexion of any crystal using (11.13), (11.14) and Table I.

The Lorentz factor is large for points close to the rotation axis, with $\xi \approx 0$. Such points are always near the sphere of reflexion: a relatively long time is taken in cutting through the sphere, and the rocking curves relating the diffracted intensity with the angular position of the crystal are correspondingly wide. Reflexions lying in a cylindrical region around the goniometer-head axis cannot be measured with any accuracy because of the large Lorentz factor; they are best examined in an alternative mounting of the crystal.

(c) *Anti-equi-inclination setting* ($\mu = \nu$). In this setting, which can only be used for the zero level ($\zeta = 0$), the incident and reflected beams are equally inclined to the rotation axis, with $\mu = \nu$. Thus $\cos\mu = \cos\nu$, and L^{-1} is given by the same equation, (11.13), as for the equi-inclination setting. In this equation (11.13) ν can have any value, provided it is equal to μ, and Υ is given by equation (2.22):

$$\sin\tfrac{1}{2}\Upsilon = \frac{\xi}{2\cos\nu}.$$

(d) *Flat-cone setting* ($\nu = 0$). The reflected beam emerges at 90° to the rotation axis, $\nu = 0$, and (11.10) reduces to

$$L^{-1} = \cos\mu \sin\Upsilon. \qquad (11.15)$$

The angles μ, Υ in this expression are related to the cylindrical co-ordinates ξ, ζ by equations (2.23) and (2.24):

$$\sin\mu = -\zeta,$$

and

$$\cos\Upsilon = \frac{2 - \xi^2 - \zeta^2}{2(1 - \zeta^2)^{\frac{1}{2}}}. \qquad (11.16)$$

11.2. Polarization factor

In X-ray diffraction, a polarization factor arises because of the dependence of the scattered amplitude on the orientation of the electric vector $\mathbf{E}$ of the X-ray beam. From the Thomson theory of X-ray scattering, the amplitude scattered by a single electron is proportional to $\sin\phi_0$, where ϕ_0 is the angle between $\mathbf{E}$ and the direction of the reflected beam. Accordingly, a factor p is introduced in the basic equation (11.3) to account for the $\sin\phi_0$ effect of polarization.

The magnitude of p depends on the degree of polarization of the X-ray beam. Characteristic radiation direct from the target is unpolarized, whereas radiation is partially polarized by reflexion at a crystal monochromator. We shall discuss, therefore, the evaluation of p separately for direct and for monochromatized incident radiation.

In neutron diffraction, there is no polarization effect associated with the nuclear reflexions, and so $p = 1$.

Incident unpolarized radiation

We resolve the incident beam into two equal components with their electric vectors parallel and perpendicular to the plane containing the incident and reflected beams (see Fig. 121). The amplitude of each reflected component is proportional to $\sin\phi_0$, where ϕ_0 is $90° - 2\theta$ for the parallel component and $90°$ for the perpendicular component. Thus the amplitudes are in the ratio $\cos 2\theta : 1$ and the intensities in the ratio $\cos^2 2\theta : 1$. The mean intensity for the two states of polarization is defined as the polarization factor p:

$$p = \tfrac{1}{2}(1 + \cos^2 2\theta). \tag{11.17}$$

For $\theta = 0$ or $90°$, p is unity and the intensities of the parallel and perpendicular components are equal, that is, the reflected beam is unpolarized. For $\theta = 45°$, $p = \tfrac{1}{2}$ and the intensity of the parallel component is zero, so that the reflected beam is completely polarized. (Chandrasekhar (1960b) used this property to produce a polarized X-ray beam in his method of correcting intensities for the effect of extinction (p. 192.) At all other values of θ the reflected X-ray beam is partially polarized (Fig. 122).

The basic equation (11.3) applies to an ideally mosaic crystal with diffracted intensities proportional to F^2. For a perfect crystal, the integrated intensities are proportional to $|F|$, not F^2, and the polarization factor in (11.17) is modified to

$$p = \tfrac{1}{2}(1 + |\cos 2\theta|). \qquad (11.18)$$

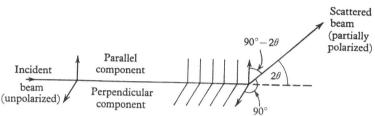

Fig. 121. Derivation of polarization factor for unpolarized incident beam.

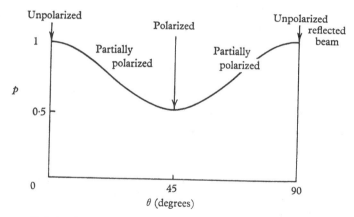

Fig. 122. Polarization factor for unpolarized incident beam, as a function of θ.

We usually assume that the specimen crystal behaves like a mosaic crystal, and express any departure from this behaviour by the term 'extinction'. Those reflexions which require an extinction correction have a polarization factor which is intermediate between (11.17) and (11.18) and is dependent on the strength of the reflexion and on the degree of perfection of the crystal. This emphasizes yet another difficulty in dealing with the problem of extinction. To derive accurate structure factors, the extinction must be small, as this not only allows an adequate estimate to be made of the extinction correction but also ensures that there is no uncertainty in the polarization factor given by equation (11.17).

Incident monochromatized radiation: equatorial geometry

If θ_M is the Bragg angle of the monochromator, the beam striking the sample is partially polarized, with the intensities of the parallel and perpendicular components in the ratio $\cos^2 2\theta_M : 1$. In the equatorial method, the plane of incidence for the sample coincides with the horizontal (equatorial) plane; if we assume that the same plane coincides with the plane of incidence for the monochromator, the polarization factor is given by

$$p = \frac{1 + \cos^2 2\theta_M \cos^2 2\theta}{1 + \cos^2 2\theta_M}. \qquad (11.19)$$

This formula is derived with the aid of Fig. 123: the intensity of the beam striking the sample is proportional to $\frac{1}{2}(1 + \cos^2 2\theta_M)$ and the intensity reflected by the sample is proportional to

$$\frac{1}{2}(1 + \cos^2 2\theta_M \cos^2 2\theta).$$

In equation (11.19) we assume that *both* the monochromator and the sample behave like ideally mosaic crystals. However, the monochromator is set to reflect radiation at a family of strongly reflecting planes, and it seems likely that it operates under conditions of strong extinction: if appreciable attenuation of the beam occurs within a single mosaic block, the monochromator approaches the behaviour of a perfect crystal. For such a crystal the polarization factor in (11.19) is replaced by

$$p = \frac{1 + |\cos 2\theta_M| \cos^2 2\theta}{1 + |\cos 2\theta_M|}. \qquad (11.20)$$

The uncertainty in the correct form of polarization factor is reduced by selecting θ_M close to 0 (or 90°), and so it is preferable to use a reflecting plane of low Bragg angle θ_M. Miyake, Togawa & Hosoya (1964) have shown that with CuKα radiation, reflected by the (200) plane of a lithium-fluoride monochromator for which

$$2\theta_M = 45°,$$

the polarization factor given by equation (11.19) had a maximum error of 3·5 per cent when θ for the specimen was 45°. They concluded that the effect of extinction in the monochromator is important for accurate intensity measurements and requires a separate examination of the monochromator to determine the correct form of p, which is intermediate between (11.19) and (11.20).

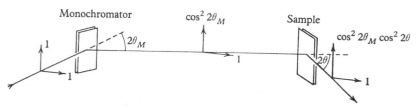

Fig. 123. Derivation of polarization factor for crystal-reflected incident beam. The primary, once-reflected and twice-reflected beams are co-planar.

Incident monochromatized radiation: inclination geometry

We shall quote without proof the expression for the polarization factor in the general inclination method using monochromatized radiation (Azaroff, 1955):

$$p = \frac{(\cos^2 2\theta_M \cos^2 \epsilon + \sin^2 \epsilon)\cos^2 2\theta + \cos^2 2\theta_M \sin^2 \epsilon + \cos^2 \epsilon}{1 + \cos^2 2\theta_M}. \quad (11.21)$$

Here θ_M is the Bragg angle of the monochromator, θ that of the sample, and ϵ is the angle between the planes of incidence at the monochromator and at the sample. It is assumed that the monochromator (and sample) reflect as ideally mosaic crystals, although we have already noted the inherent danger in this assumption.

To apply equation (11.21) to a particular type of setting we must evaluate ϵ in terms of the setting angles. For the normal-beam setting, ϵ is given by (Levy & Ellison, 1960):

$$\left.\begin{array}{l} \sin\epsilon = \sin\nu\operatorname{cosec}2\theta, \quad \sigma = 90°, \\ \cos\epsilon = \sin\nu\operatorname{cosec}2\theta, \quad \sigma = 0, \end{array}\right\} \quad (11.22)$$

where σ is the angle between the plane of incidence at the monochromator and the plane generated by the Weissenberg rotation axis and the primary monochromatized beam. For the equi-inclination setting, ϵ is given by:

$$\left.\begin{array}{ll} \sin \epsilon = \tan \nu \cot \theta, & \sigma = 90°, \\ \cos \epsilon = \tan \nu \cot \theta, & \sigma = 0. \end{array}\right\} \qquad (11.23)$$

The angle σ is constant for a particular experimental configuration of monochromator and diffractometer.

11.3. Observed structure factors: absolute measurements

Nowadays it is customary to base crystal-structure determinations on relative structure factors, although the importance of absolute intensities in structural work has been emphasized by Bragg & West (1928) and more recently by Lipson & Cochran (1957). If the scale factor c in (11.3) is known, it is easier to detect the presence of systematic errors in the intensity data: extinction, for example, reduces the observed intensities of the strongest reflexions below the values calculated for an ideally mosaic crystal, and the presence of extinction is revealed by comparing the experimental value of c with that determined by a least-squares comparison of the observed and calculated intensities. Accurate bond distances and angles require a proper correction for the effects of thermal motion (Cruickshank, 1956): errors in the thermal parameters are caused by scale factor errors, and it is preferable to measure c independently so as to avoid having to introduce it as an adjustable parameter in the analysis of the intensity data. In the study of defect structures the number of atoms in the 'average unit cell' is not integral, but can be determined from absolute intensity measurements. Absolute values are also important in the study of non-Bragg scattering, such as thermal diffuse scattering or defect diffuse scattering, and in the application of direct methods to the solution of the phase problem. Finally, from a knowledge of the absolute magnitudes of the structure factors, we can calculate the ratio of the observed strength of the reflexions to the hypothetical strength with all atoms scattering in-phase: this information may be of considerable significance in the subsequent determination of the structure.

An approximate value of c can be derived from a complete set of relative intensities ρ_{hkl} without further recourse to experiment: the

ρ_{hkl}'s contain their own inherent absolute standard. Wilson's method (Wilson, 1942) is based on the statistical result that the mean value of $|F_{hkl}|^2$ is equal to the sum of the squares of the scattering factors f_j for all the N atoms in the unit cell:

$$\langle|F_{hkl}|^2\rangle = \sum_{j=1}^{N} f_j^2. \tag{11.24}$$

This equation follows directly from the expression for F_{hkl} in terms of the atomic positions $(x_j y_j z_j)$ in the cell

$$F_{hkl} = \sum_{j} f_j e^{2\pi i(hx_j+ky_j+lz_j)}, \tag{11.25}$$

assuming that there is a large number of atoms in the unit cell, with none occupying positions of special symmetry. The value of f_j in (11.24) must be corrected for the effect of thermal motion by introducing an approximate correction of the form $e^{-\bar{B}\sin^2\theta/\lambda^2}$, where $\bar{B}$ is the 'overall temperature factor'. This exponential term is constant for reflexions in a sufficiently narrow range of $\sin\theta/\lambda$ and is unity for $\sin\theta/\lambda = 0$. Thus the mean relative values of $|F^{\text{obs.}}|^2$ can be evaluated over small ranges of $\sin\theta/\lambda$, and the constants c' within each range found from the equation

$$c'\langle|F_{hkl}^{\text{obs.}}|^2\rangle = \sum_{j}(f_j')^2,$$

where f_j' is the theoretical scattering factor, uncorrected for thermal motion. If $\ln c'$ is plotted against the mean value of $\sin^2\theta/\lambda^2$ in each range, an approximately linear relationship is obtained and the intercept at $\sin\theta/\lambda = 0$ gives the required value of $\ln c$.

The statistical result (11.24) will not apply strictly if there are atoms in special symmetry positions or if the number of reflexions in a given range of $\sin\theta/\lambda$ is small, and an error of 30 per cent in c is typical of Wilson's method. Better values, correct to 10 per cent, are given by Kartha's method (Kartha, 1953), which uses an exact relationship between the sum of the squares of the observed structure factors and an integral involving the atomic scattering factors. Macintyre (1963) describes the application of this relationship in the routine reduction of the intensity data to absolute structure factors.

19

Experimental determination of scale factor

The scale factor can be more precisely determined by experimental methods. The simplest procedure involves comparing the intensities of the Bragg reflexions from the sample with those from a standard crystal, examined under identical conditions. On p. 239 we noted the importance of knowing exactly the quantity μR in calculating the *absolute* magnitude of the absorption factor of a spherical crystal, where R is the radius of the crystal and μ its coefficient of linear absorption. The absorption in both the sample and standard crystals should be very small, and so the method is less satisfactory for X-rays than for neutrons.

Other experimental methods used in X-ray work are based on the measurement of the intensity of the main beam striking the sample. From equations (9.1) and (11.1)

$$\frac{E\omega}{I_0} = N_c^2 \lambda^3 |F_{hkl}|^2 \left(\frac{e^2}{mc^2}\right)^2 Lp\,\delta V, \qquad (11.26)$$

where E is the reflected energy in the hkl reflexion, ω is the angular velocity of the crystal, I_0 is the intensity (energy per unit area per unit time) of the main beam, N_c is the number of unit cells per unit volume, and δV is the volume of the crystal. In making relative intensity measurements we equate the reflected energy, corrected for absorption, to $Lp|F_{hkl}|^2$, so that the scale factor is

$$c = \frac{E}{Lp|F_{hkl}|^2} = \frac{N_c^2 \lambda^3 \left(\dfrac{e^2}{mc^2}\right)^2 \delta V I_0}{\omega}. \qquad (11.27)$$

c can be readily found from the quantities on the right-hand side of (11.27), provided I_0 is known.

The main X-ray beam is too strong to be measured directly with a proportional counter: it must be attenuated first so that the dead-time of the counter is appreciably less than the mean time interval between the arrival of the X-ray quanta. To evaluate the intensity I_0, the attenuation factor must be accurately known. Multiple foils can be used for attenuation, but difficulties are caused by the progressive hardening of the beam, due to the preferential absorption of the softer components in the passage through successive foils. The multiple-foil technique is, therefore, most suitable for measurements with monochromatized radiation or with balanced filters: according to Burbank (1965), it is then capable of establishing the intensity scale factor to within 3 per cent of its true value. Wagner, Witte & Wölfel (1955) compared the flux in the direct and diffracted beams with an ionization chamber. They attenuated the direct beam by means of a rotating sector: this method avoids changes in the spectral composition of the beam caused by filtering, but it can only be used with a current ionization chamber which has no counting losses. Another method of determining I_0 (Buyers, 1964) consists of measuring the intensity scattered by a small block of paraffin placed in the main beam, and using theoretical values for the absolute magnitude of this scattering. In an amorphous solid such as paraffin the scattering from each atom bears a random phase relation to that from the other atoms; thus the total scattering is contributed by independent unmodified scattering and by modified (Compton) scattering, and both of these can be calculated from the unmodified and modified scattering intensities of the separate atoms.

11.4. Calculated structure factors

In §11.5 we shall examine the experimental measurements of the structure factors of a number of crystals and discuss these measurements in relation to the magnitude of the R-index, defined by

$$R = \sum_{hkl} \left| \left| F_{hkl}^{\text{obs.}} \right| - \left| F_{hkl}^{\text{calc.}} \right| \right| \div \sum \left| F_{hkl}^{\text{obs.}} \right|. \tag{11.28}$$

If the structure is known, that is, the positions of the atoms in the unit cell are known, this index gives some indication of the errors in the observed structure factors, *provided* the $F_{hkl}^{\text{calc.}}$ terms are known precisely: R is zero, only if there are no uncertainties in the observed *and* calculated structure factors. Thus, first we must discuss the limitations in evaluating the calculated structure factors.

The calculated structure factors are given by

$$F_{hkl}^{\text{calc.}} = \sum_{j=1}^{N} f_j e^{2\pi i(hx_j + ky_j + lz_j)}. \tag{11.29}$$

The same formula applies in neutron diffraction with the X-ray atomic scattering factor f_j replaced by the nuclear scattering amplitude b_j. $x_j y_j z_j$ are the co-ordinates of the jth atom in the unit cell, expressed as fractions of the cell edges, and the summation extends over all N atoms in the cell. f_j must be computed for the atom undergoing thermal motion: the effect of this motion is allowed for by writing f_j as the product of two quantities

$$f_j = f_{0,j} \times T_j, \qquad (11.30)$$

where $f_{0,j}$ is the value of f_j for the atom at rest and T_j is a temperature correction factor dependent on the particular atom j. The evaluation of these two quantities may lead to uncertainties in determining $F^{\text{calc.}}$, even though the $x_j y_j z_j$'s are exactly known.

Atomic scattering factor for stationary atom

The X-ray scattering factor $f_{0,j}$ (or f_0) is defined as the ratio of the amplitude of the radiation scattered by the atom at rest to the amplitude scattered under the same conditions by an electron. The electrons in the atom occupy a volume whose linear dimensions are comparable with the wavelength of X-radiation, so that the phase differences between X-rays scattered by different parts of the atom must be taken into account in evaluating f_0. At low angles of diffraction these phase differences are small and the value of f_0 is simply the total number of electrons in the atom or ion. As the scattering angle 2θ increases, the scattering amplitude is reduced by interference: a typical theoretical curve showing this form-factor dependence on $\sin \theta / \lambda$ is shown in Fig. 124.

The calculation of the X-ray scattering factors of different atoms and ions is described by James (1962) and in volume III of *The International Tables*. The results are presented in the form of tables giving f_0 at fixed intervals of $\sin \theta / \lambda$, and the scattering factor at any value of $\sin \theta / \lambda$ is readily found by interpolation. These calculated scattering factors are reliable only in so far as the total wave function used in the calculations is a reliable representation of the electron density: the exact form of wave function is not known for any atom, with the exception of hydrogen. Furthermore, the calculations usually assume that the electron density of the atom is spherically symmetrical, whereas many atoms are aspherical and the scattering factor is a function not only of $\sin\theta/\lambda$ but also of the orientation of the atom with respect to the direction of the reflected beam.

The validity of scattering factors based on a spherically averaged electron density (the 'spherical approximation') has been examined by Dawson (1964a, b). Dawson considers the example of bonding

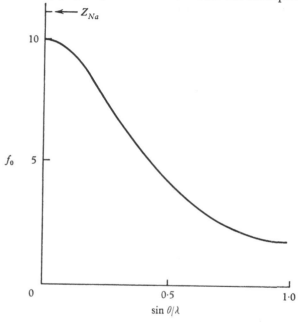

Fig. 124. Atomic scattering factor for Na^+ at rest, plotted against $\sin\theta/\lambda$. The charge distribution is spherical and there is no acentric component of f_0 (cf. Fig. 125).

electrons which impose an aspherical charge distribution on the scattering atom; he finds that for non-centro-symmetric valency states the scattering factor is complex and of the form

$$f_0 = f_c + if_a,$$

where f_c corresponds to the centric component of the overall charge distribution and f_a to the antisymmetric component. f_a is zero at $\theta = 0$, and the general form of f_c and f_a as a function of $\sin\theta/\lambda$ is illustrated in Fig. 125. The antisymmetric component f_a, which is 90° out-of-phase with f_c, is ignored in the spherical approximation.

Dawson has calculated that for a simple hypothetical structure, consisting of two sp^3 nitrogen atoms which are centrosymmetrically disposed in a unit cell, the contribution to the R-index arising from the neglect of the intrinsic shape of the nitrogen atoms is as high as 8 per cent. This R-index would be obtained by comparing a perfect set of observed structure factors with a set of structure factors calculated on the basis of the spherical approximation. Careful experimental work on a number of simple structures is

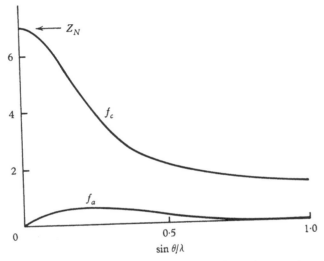

Fig. 125. Centric f_c and acentric f_a components of atomic scattering curve of nitrogen in sp^3 valence state (after Dawson, 1964b).

required before Dawson's work can be fully evaluated, but it is clear that the spherical approximation can be a serious limitation to the calculation of accurate X-ray scattering factors of stationary atoms.

In neutron diffraction, the scattering centre is the nucleus of the atom. The radius of the nucleus is approximately 10^{-12} cm, so that it behaves as a point scatterer for wavelengths of the order of 1 Å. For this reason, there is neither a $\sin\theta/\lambda$ dependence nor an orientation dependence of the atomic scattering amplitude. The scattering amplitude b for the atom at rest of all but a few elements is independent of the wavelength λ: it has the dimensions of length and is usually expressed in units of 10^{-12} cm. In the absence of an adequate theory of the nucleus, b is measured experimentally: it is tabulated for different elements and isotopes by Bacon (1962) and in many cases is known to an accuracy of better than 1 per cent.

Thermal motion

To take into account the effect of thermal motion, the correction factor T_j in equation (11.30) must be determined for each atom j. For isotropic thermal motion, T_j can be written as

$$T_j = e^{-B_j \sin^2 \theta / \lambda^2}, \quad (11.31)$$

where B_j is the 'isotropic temperature factor'. B_j is a constant for each atom at a given temperature; it is related to the mean-square displacement $\overline{u^2}$ of the atom in any direction from its mean position by the equation

$$B = 8\pi^2 \overline{u^2}. \quad (11.32)$$

B is usually treated as an empirical constant to be derived from a least-squares comparison of the observed and calculated intensities.

In the more general case of anisotropic vibration of the atom j, equation (11.31) for T_j is replaced by (Cruickshank, 1956):

$$T_j = \exp[-(b_{11}h^2 + b_{12}hk + b_{13}hl + b_{22}k^2 + b_{23}kl + b_{33}l^2)]. \quad (11.33)$$

The anisotropic thermal motion is represented by an ellipsoid of vibration in reciprocal space: six temperature factors b_{11}, b_{12}, ... for each atom define the principal axes and direction cosines of the ellipsoid. (This number may be reduced by symmetry for atoms at special positions.) As in the isotropic case, the temperature factors are treated as adjustable constants in the least-squares refinement of the experimental data.

It is important to realize that both (11.31) and (11.33) are valid only within the limitations of the 'harmonic approximation'. Thermal motion causes the atoms in a crystal to execute small oscillations about their mean positions: the harmonic approximation means that, in expanding the potential energy of the crystal in powers of the amplitudes of these oscillations, all terms beyond those which are quadratic in the amplitudes are neglected. Harmonic interatomic forces imply that the crystal has no thermal expansion and has other properties not possessed by real crystals.

The harmonic approximation also requires that the so-called 'smearing functions' around atom centres are ellipsoids (transforming into ellipsoids in reciprocal space), which degenerate into spheres for atoms at sites of cubic point symmetry: anharmonicities cause the smearing functions to be more complicated, although they must still conform with the local site symmetry. Experimental diffraction measurements which demonstrate the breakdown of the harmonic approximation have been described by Willis (1965).

We can summarize this section by stating that the evaluation of $F^{\text{calc.}}$ in equation (11.29) cannot be made exactly. An estimate of the correction due to thermal motion, using up to six adjustable parameters to describe the vibration ellipsoid of each atom, ignores the effect of anharmonicities. There are serious limitations to current methods of calculating the X-ray scattering factor of a stationary atom, although there is no corresponding difficulty in describing the nuclear scattering amplitude in neutron diffraction.

11.5. Comparison of observed and calculated structure factors

The need for improving the accuracy of structure determinations has been stressed by Cruickshank (1960): if bond lengths are measured to within a limit of error of 0·01 Å, the final R-index must approach a value of about 1 per cent. We have noted that a low R-index requires not only accurate intensity measurements, corrected for absorption, extinction and other systematic errors, but also accurate theoretical scattering factors, corrected for anomalous dispersion, orientation effects and thermal motion. We shall describe now some results obtained with X-ray and neutron diffractometers on a few crystals of known structure.

Rock-salt (X-rays)

Some X-ray measurements by Abrahams (1964) are listed in Table XXVIII. These were carried out on a small sphere of sodium chloride, using an equi-inclination diffractometer with beta-filtered CuKα radiation. All equivalent reflexions accessible to the diffractometer were measured: the number of equivalent reflexions of the general type {hkl} is 48, and this large redundancy allows the use of statistical methods in estimating the mean value of the hkl structure factor $F^{\text{obs.}}$ and the variance s^2 of the mean. The last column in Table XXVIII represents the standard deviation σ derived from counting statistics alone (see p. 269). Finally, the $F^{\text{calc.}}$ values were derived from the equations:

$$F^{\text{calc.}} = 4f_{\text{Na}}\exp\left(-B_{\text{Na}}\sin^2\theta/\lambda^2\right) + 4f_{\text{Cl}}\exp\left(-B_{\text{Cl}}\sin^2\theta/\lambda^2\right)$$
$$\ldots h, k, l \text{ even,}$$

$$F^{\text{calc.}} = 4f_{\text{Na}}\exp\left(-B_{\text{Na}}\sin^2\theta/\lambda^2\right) - 4f_{\text{Cl}}\exp\left(-B_{\text{Cl}}\sin^2\theta/\lambda^2\right)$$
$$\ldots h, k, l \text{ odd,}$$

$$(11.34)$$

where f_{Na} and f_{Cl} are the atomic scattering factors for Na^+ and Cl^-, and B_{Na} and B_{Cl} the individual atomic temperature factors. (These equations follow from (11.29), with four sodium atoms at $000 +$ face-centred positions and four chlorine atoms at $00\frac{1}{2} +$ face-centred positions: isotropic temperature factors are used, as all the atoms occupy sites of cubic point symmetry, $m3m$.) The intensity measurements were made on a relative scale, and the scale factor, together with the temperature factors B_{Na}, B_{Cl}, were treated as adjustable parameters in the least-squares refinement of the structure factor data.

The table shows that the fractional error in $F^{\text{obs.}}$, when expressed as $s/F^{\text{obs.}}$, varies from less than 1 per cent for the strong, even-index reflexions to about 6 per cent for the weaker, odd-index reflexions. Only a small part of this error is explained by counting statistics: σ is, on average, only one-seventh of s. Clearly, the random errors in $F^{\text{obs.}}$ have been reduced to such an extent that the principal errors are of systematic nature.

If we assume that different estimates of $F^{\text{obs.}}$ would be normally distributed about the 'true' value, the mean deviation of each estimate is

$$\sqrt{\frac{2}{\pi}}s = 0.7979s.$$

Thus the contribution to the R-index due to the uncertainty in $F^{\text{obs.}}$ is

$$\sum_{hkl} 0.7979 s_{hkl} \div \sum_{hkl} |F^{\text{obs.}}|,$$

which is 1·3 per cent for the figures in Table XXVIII. This is not sufficient to account for the actual value for the R-index of 2·1 per cent. It is unlikely that the extra 0·8 per cent can be ascribed to uncertainties in estimating $F^{calc.}$: from the discussion in §11.4, appreciable errors in $F^{calc.}$ can arise if the atoms are aspherical or execute large amplitudes of thermal vibration, but neither of these conditions applies to NaCl. We conclude that the estimated standard deviation s, determined from the variance in the measurements of symmetry-equivalent reflexions, is probably an underestimate of the real standard deviation, but the origin of this difference is not known.

TABLE XXVIII. *Sodium chloride structure factors at* 295° *K* (*X-rays*) (*Abrahams*, 1964)

hkl	$\overline{F^{obs.}}$	$F^{calc.}$	$\overline{F^{obs.}} - F^{calc.}$	s	σ
111	19·84	19·79	+0·05	0·18	0·06
200	79·34	83·43	−4·09	0·55	0·05
220	66·96	67·03	−0·07	0·40	0·06
311	10·90	10·08	+0·82	0·44	0·13
222	57·39	56·73	+0·66	0·31	0·05
400	50·17	49·55	+0·62	0·25	0·04
331	9·86	9·56	+0·30	0·36	0·13
420	44·97	44·16	+0·81	0·27	0·08
422	39·99	40·03	−0·04	0·34	0·08
333	10·00	9·99	+0·01	0·38	0·13
511	10·11	9·99	+0·12	0·35	0·11
440	32·78	33·23	−0·45	0·17	0·08
531	9·76	10·10	−0·34	0·53	0·12
442	30·47	30·62	−0·15	1·61	0·08
600	30·42	30·62	−0·20	1·22	0·05
620	28·76	28·24	+0·52	0·63	0·07
533	9·15	9·87	−0·72	0·67	0·09
622	24·18	26·07	−1·89	0·35	0·05

In this table: s^2 is variance of $\overline{F^{obs.}}$, estimated from measurements on symmetry-equivalent reflexions; σ is standard deviation from counting statistics; $F^{calc.}$ is calculated from $B_{Na} = 1·60$ Å², $B_{Cl} = 1·17$ Å².

Some further measurements of Abrahams (1965) using an equi-inclination diffractometer with balanced filters indicate that the standard errors of the observations on NaCl are up to 4·0 per cent. Half of this is attributed to instrumental causes and half to systematic errors (absorption, extinction, etc.).

Tetrahedrite (X-rays)

The reliability of X-ray equi-inclination counter-diffractometer data has also been examined by Wuensch (1963), from the analysis of data recorded on the mineral tetrahedrite, $Cu_{12}Sb_4S_{13}$. Diffracted intensities were obtained from a spherical sample using β-filtered CuKα radiation. Tetrahedrite is cubic, space-group I $\overline{4}3m$, and each general reflexion occurs as 48 symmetry-equivalent reflexions. Thus, as with NaCl, the comparison of reflexions required by symmetry to be equivalent provides a method of checking the data. Three kinds of check were carried out:

(1) Repeated measurement of certain reflexions, over periods of several weeks. The reproducibility of these measurements is affected by counting statistics and by drift in the electronic equipment.

(2) Comparison of equivalent reflexions, hkl and khl, within a given level l. Systematic errors such as absorption affect the agreement.

(3) Comparison of equivalent reflexions, hkl and hlk, between different levels. Agreement is affected by further systematic errors, such as differences in the way in which the white-radiation streak is crossed.

Some of the results are summarized in Table XXIX. The deviation between equivalent structure factors both within each level and between levels exceeds the reproducibility of the data. This difference must be ascribed to systematic errors in measuring $\bar{F}$. The main source of error was absorption, and the uncertainty in the absorption correction (arising from a 1·7 per cent variation in radius of the nominally spherical sample) was sufficient to account for the increase in the r.m.s. deviation in tests (2) and (3).

TABLE XXIX. *Reliability of structure factors for cubic tetrahedrite (X-rays) (Wuensch, 1963)*

Test	Range of deviations from mean F (%)	r.m.s. deviation (%)
(1) Reproducibility	0·016–2·8	1·15
(2) Comparison of hkl and khl	0·066–7·5	2·92
(3) Comparison of hkl and hlk	0·43 –6·0	2·82

Uranium dioxide (neutrons)

A systematic study of the neutron reflexions of a spherical crystal of cubic UO_2 has been undertaken by Rouse & Willis

(1966). The reflexions were recorded at $\lambda = 1 \cdot 04$ Å using an automatic four-circle diffractometer.

Table XXX summarizes their results. $\overline{F^{\text{obs.}}}$ in the third column is the mean of the measurements on n symmetry-related reflexions, where n is the number in the second column. S^2 is the variance of the n measurements, given by the equation

$$S^2 = \sum_{i=1}^{n} \frac{(F_i - \overline{F})^2}{(n-1)},$$

and σ^2 is the variance expected from Poisson counting statistics. The reflexions occur in three groups, strong, medium and weak, corresponding to $h+k+l = 4m$, $4m \pm 1$ and $4m+2$, respectively (m = integer). The variance S^2 is related to the variance s^2 of $\overline{F^{\text{obs.}}}$ by

$$s^2 = \frac{S^2}{n}$$

(see, for example, *Statistics* section in *International Tables for X-ray Crystallography*, vol. II).

A measure of the consistency of the internal (S) and external (σ) errors of a *single* observation is provided by the last column of Table XXX. The quantity $\chi^2 = (n-1)S^2/\sigma^2$ is distributed in the χ^2-distribution on $n-1$ degrees of freedom, and P_{χ^2} is the probability that χ^2 for the sample exceeds that actually found. If P_{χ^2} is close to $0 \cdot 5$ there is no inconsistency between the internal and external estimates of error, and if P_{χ^2} is near zero or unity the estimates are inconsistent and some kind of systematic error is present. Judged from this criterion, the weak reflexions are most consistent, the medium reflexions are less so, and the strong reflexions are the worst.

The $F^{\text{calc.}}$ values for the three groups of reflexions are given by the equations

$$
\begin{aligned}
F^{\text{calc.}} &= 4b_U e^{-B_U \sin^2 \theta/\lambda^2} + 8b_O e^{-B_O \sin^2 \theta/\lambda^2} && (h+k+l = 4m) \\
&= 4b_U e^{-B_U \sin^2 \theta/\lambda^2} && (h+k+l = 4m \pm 1) \\
&= 4b_U e^{-B_U \sin^2 \theta/\lambda^2} && \\
&\quad - 8b_O e^{-B_O \sin^2 \theta/\lambda^2} && (h+k+l = 4m+2).
\end{aligned}
$$

Here b_U, b_O are the nuclear scattering amplitudes of uranium and oxygen, and B_U, B_O are the two temperature factors. Preliminary analysis of the data in Table XXX using these equations showed that the strong reflexions were affected by extinction and so these reflexions were omitted from the final least-squares refinement of the remaining 14 reflexions. The final analysis gave $B_U = 0.19$ Å², $B_O = 0.43$ Å², $b_U/b_O = 1.47$ and an R-index of 0.9 per cent. This

TABLE XXX. *Uranium dioxide structure factors (neutrons)*

	{hkl}	No. of independent observations n	$\overline{F^{obs.}}$	S	σ	$S/\overline{F^{obs.}}$ (%)	P_{χ^2}
Strong reflexions	008	4	100·44	0·51	0·24	0·5	<0·01
$(h+k+l = 4m)$	022	10	115·34	2·31	0·23	2·0	<0·01
	026	17	108·66	1·32	0·24	1·2	<0·01
	044	7	107·28	0·77	0·24	0·7	<0·01
	224	14	113·32	0·86	0·24	0·8	<0·01
	246	25	104·16	0·80	0·23	0·8	<0·01
	444	3	105·58	1·26	0·23	1·2	<0·01
Medium reflexions	113	14	52·65	0·40	0·24	0·8	<0·01
$(h+k+l = 4m \pm 1)$	115	15	51·24	0·37	0·24	0·7	<0·01
	117	14	49·31	0·29	0·24	0·6	0·13
	133	12	52·02	0·36	0·24	0·7	0·02
	135	27	50·62	0·35	0·25	0·7	<0·01
	137	27	48·60	0·39	0·25	0·8	<0·01
	155	12	49·15	0·30	0·24	0·6	0·10
	333	6	51·52	0·13	0·24	0·3	0·92
	335	12	49·86	0·32	0·25	0·6	0·07
	553	12	48·88	0·26	0·24	0·5	0·30
Weak reflexions	006	5	14·94	0·37	0·34	2·5	0·32
$(h+k+l = 4m+2)$	024	18	17·62	0·32	0·33	1·8	0·52
	046	16	12·86	0·34	0·34	2·6	0·49
	244	13	15·31	0·27	0·32	1·8	0·73

R-index was reduced even further, to 0·5 per cent, by introducing one extra parameter to represent the anharmonic contribution to B_O.

It is too early to draw any general conclusions about the accuracy of X-ray and neutron diffractometer data. From the very limited studies to date, it seems that a final R-index approaching 1 per cent is attainable in cubic crystals provided measurements are averaged over symmetry-equivalent reflexions. Furthermore, there is some evidence that a lower R-index is obtained with neutrons than with X-rays: this may be due to the lower absorption correction with neutrons and to the difficulty of calculating X-ray atomic scattering factors.

CHAPTER 12

COMPUTER PROGRAMS AND ON-LINE CONTROL

12.1. Diffractometer input and output

The automatic diffractometer, whatever its type, can be regarded as a black box which accepts numerical input information of one kind and produces numerical information of another kind. The input information consists of the settings of the various crystal and detector shafts and all the instructions necessary to make valid measurements of the intensities of Bragg reflexions. The output data consist of the results of those measurements. The generation of the input information and the processing of the diffractometer output must inevitably be carried out by means of a computer: the rate at which measurements are made by an automatic diffracto-meter is such as to make manual computation with a desk calcu-lator quite impracticable.

It is important, therefore, that the input and output medium of an automatic diffractometer be one which is readily generated and read by a computer. (In the case of analogue diffractometers, discussed in §3.3, only the output medium needs to be considered.) The most common media are punched cards or punched paper tape: the choice between these two is generally dictated by the computer installation which is used in conjunction with the diffractometer. It may be noted in passing that punched cards offer a greater flexibility in that the sequence in which reflexions are to be measured and the sequence in which experimental results are listed can be varied by simply resorting a stack of cards. The total number of cards which can be read or punched without manual intervention is, however, limited by the capacity of the card-holding hoppers of commercially available punches and readers. These hoppers cannot hold more than 1,000 cards: card-controlled diffractometers usually require at least one input card per reflexion and produce at least one output card. The hoppers are loaded with alternate input and blank cards and so the maximum number of reflexions which can be measured before reloading is 500.

In contrast, punched paper tape is less versatile but the maximum length of a reel of tape is virtually unlimited (a standard 1,000 ft reel of tape can store about 10^5 characters). Tape is mostly used for diffractometers with an incremental shaft-setting system in which the sequence of reflexions cannot be varied in any case.

There is little national, let alone international, standardization of tape codes. A number of 5-, 6-, 7- or 8-holes-per-character codes are in use with different computers. Fortunately, it is normally quite easy to adapt diffractometer input and output circuits to any desired code.

Both cards and tape suffer from two basic disadvantages. The equipment for handling them is relatively unreliable compared with modern solid-state electronic circuits: faults in the input and output equipment are the most common causes of interruptions in the functioning of a diffractometer installation. In addition, these media can only be punched and read at relatively slow speeds: punching and reading speeds do not normally exceed 100 and 1,000 characters per second, respectively. Both these faults are aggravated with output equipment which produces an output in printed characters at the same time as punching the information. An automatic diffractometer, particularly an X-ray diffractometer, can give rise to a prodigious quantity of numerical output: the two diffractometers described in §§3.8 and 3.9 can each produce more than 1.5×10^5 characters in a 24 h run. This amount of information can be read into a computer in about 10 min: the cost of this time on a large computer is appreciable. (The transmission of this information on Telex-type data-links to a distant computer takes some hours!)

Attempts are now being made to eliminate the intermediate punched paper tape or cards completely and to connect the diffractometer output directly to the input channel of an on-line computer. The prime function of the computer is then one of preliminary data reduction; in addition, it makes possible 'closed-loop' operation whose benefits are discussed below.

12.2. Types of computer programs

Three main types of computer programs will normally be required for 'off-line' or 'open-loop' operation of an automatic diffractometer. These are:

(a) shaft setting programs;

(b) data-verification programs;

(c) data-reduction programs.

For 'on-line' or 'closed-loop' operation, that is, if the diffractometer is directly connected to a controlling computer, these main programs are still required, but the verification programs, which apply various tests for the validity of the recorded measurements, become much more important. In 'open-loop' operation faulty results are simply detected and listed; in 'closed-loop' operation various fault-correcting routines may be invoked to ensure an error-free remeasurement of the reflexion in question.

(a) Setting programs

The precise nature of the setting program depends very much on the geometrical arrangement of the diffractometer and on the format of its input instruction code. Analogue diffractometers require no input program as such, but all digitally set instruments must have input programs which are similar in their general form. Table XXXI shows the initial information which is required for a setting program.

The formulae for computing the setting angles are given in Chapter 2. The time taken to produce a diffractometer input tape or deck of cards depends much more on the required format of this input than on the time taken to perform the actual computations, even on a computer which uses sub-routines for generating trigonometrical functions.

When a small computer is used to generate the settings, it is often desirable to write the programs in machine code rather than in autocode in order to economize in the amount of computer core storage. An example of an actual program may serve as a very rough indication of the amount of core store required. A general setting program for three- and four-circle diffractometers has been written in the Laboratory of Molecular Biology, Cambridge, for the

Argus 304 computer. This is a fixed-point machine with a 24-bit word length: the program, which is in machine code, occupies about one-half of the 4,000-word store.

TABLE XXXI. *Initial information required for generating diffractometer input*

1. Unit cell parameters
2. Space group absences
3. Initial orientation of crystal

Normally the crystal is brought to a standard initial orientation; with on-line computers, or if there is rapid access to an off-line computer, it may be preferable to dispense with a goniometer-head, to determine the initial orientation from a small number of reflexions of known indices, and to refine the unit cell parameters and the initial orientation by a least squares method (p. 51)

4. Blind regions of diffractometer

E.g. those produced by shadowing by the χ-circle (p. 49). A maximum θ value may be specified here

5. Other restrictions on permissible setting angles

E.g. avoidance of azimuthal angles which lead to simultaneous reflexions (p. 254)

6. Sequence in which reflexions are to be measured

E.g. specify fastest moving index when the program is to permute the indices, or specify a sequence which will minimize the shaft increments between reflexions

7. List of Miller Indices

These are required if they are not generated by the program: it is always useful to have the facility of supplying a list, e.g. for remeasurements in case of errors.

8. Number and frequency of insertion of reference reflexions

9. Additional parameters under program control

E.g. oscillation range, type of scan, time spent on each reflexion, insertion of filters, etc.

(b) Data verification programs

The measurements made by the diffractometer may be invalid for a variety of reasons. Some of these are listed in Table XXXII. Before the intensity recorded by the instrument is accepted as valid and suitable for further processing it is necessary to test the results; the data-verification program will list for each reflexion either the background-corrected intensity, together with its statistical precision, or the reason why the measurement is rejected. A typical program of this kind (North, 1964), uses about

800 24-bit *Argus* locations. The list of rejected measurements may be used to generate a new setting program, if necessary with certain parameters changed, for a subsequent re-measurement.

TABLE XXXII. *Errors in experimental measurements*

Error	Test
1. Malfunctioning of output device	(a) Format of output (b) Parity of each character
2. Shafts not correctly set because of fault in setting circuits	Punch out the actual position reached by each shaft, e.g. by using an independent digitizer
3. Shafts not correctly set because of crystal slipping or incorrect unit cell parameters	(a) Reflexion should be symmetrical with respect to oscillation range (b) Reference reflexions should remain constant
4. Electrical interference in counting chain	(a) Ordinates of intensity profile should be on smooth curve (b) Results of several measurements of reflexion should agree
5. Radiation damage of crystal	Reference reflexions
6. Insufficient statistical precision of background-corrected intensity	Standard deviation of result, calculated from formulae for Poisson statistics (p. 269)

The degree of elaboration of data-verification programs will depend on the required accuracy of the results. The type of checks carried out and their frequency will be determined by actual experience of the reliability of various components of the diffracto-meter system. Tests which are not strictly necessary may decrease rather than increase the reliability of the system: the more comprehensive a verification program, the greater will be the amount of information which has to be supplied by the diffractometer. Thus, for example, test 4a of Table XXXII requires the output, to an adequate statistical precision, of a large number of ordinates for each intensity profile. This will not only slow down the rate of data collection, but will also lead to such a large output that a rapid visual inspection of the results prior to computer processing becomes impossible.

(c) *Data reduction programs*

The end-product of the data-verification program described in the paragraphs above is a list of background-corrected intensities.

These must now be converted into structure factors. The basic steps have been discussed in previous chapters; they are summarized in Table XXXIII.

TABLE XXXIII. *Data-Processing Program*

Operation	Method
1. Counting loss correction	See p. 145
2. Correction for drift of electronics or for radiation damage	Interpolation, using reference reflexions
3. Reduction of all intensities to same measuring period and filter factor	
4. Absorption and extinction corrections	See pp. 234, 243
5. Lorentz factor	See p. 278
6. Polarization factor	See p. 284
7. Scaling, combination of data from different crystals or with radiations of different wavelengths	
8. Sorting	

Not all these corrections will be carried out in every case. Thus the first operation is necessarily approximate: the best way of dealing with counting losses is to avoid them, by reducing the incident intensity when necessary. Corrections for drift or radiation damage, equally, are uncertain. Unless one is dealing with an irreplaceable crystal, if reference reflexions have undergone a progressive change by far the wisest course is to reject the results of that particular series of measurements. Absorption corrections are expensive in computer time: the necessary computation may have to be carried out subsequently on a large fast computer.

It is frequently convenient to carry out the data verification and initial processing on a small local computer; these steps lead to a great reduction in the total number of bits of information which must be stored. Subsequent computations which utilize the data, such as the calculation of Patterson syntheses, are often best carried out on a larger and more powerful computer.

12.3. Direct connexion to a computer

The concept of a small local computer used for data reduction leads naturally to the idea of the computer being connected directly to the input and output channels of the diffractometer. In the simplest version of such an arrangement the diffractometer control and measuring circuits are retained: the computer input and output simply replace the diffractometer tape or card punch, and the tape or card reader, respectively. In some recent installations much of the special-purpose control logic of the diffractometer is replaced by a general purpose computer: the only circuits specific to the diffractometer are then those associated with driving the motors and with the radiation detector. This can lead to a further simplification and increase of reliability; for example, encoding and decoding circuits do not need to be duplicated, and numbers which are to be interpreted by a computer rather than by a human operator can be in binary instead of decimal form: binary counters of the same capacity as corresponding decimal counters require fewer electronic components.

Computers are constructed in larger numbers than special-purpose hard-ware. Accordingly, computer-controlled installations should, before long, become cheaper, more reliable and easier to maintain and to service than installations which use specially designed control circuits.

Closed-loop operation

The computer can be used to adjust the measuring conditions in the light of previous measurements. The following are some examples of the situations with which a computer can deal.

(1) If a reflexion has been measured to an insufficient statistical precision the computer can initiate repetitions of the measurement until the required precision is achieved (p. 276).

(2) If the counting rate at the peak of the reflexion is so high as to lead to severe counting losses the measurement can be repeated with attenuating foils.

(3) The angular scan of the reflexion can be adjusted in the light of the measured width of the reflexion.

(4) The ratio of the time spent on the scan across the Bragg peak and on the background can be made such as to ensure the best statistical precision in a given time (p. 273).

(5) Measurements can be interrupted if the reference reflexions change in intensity.

(6) The exact shaft settings can be adjusted during the experi-

mental run to correct for inaccuracy in unit cell parameters or for small crystal movements. In this way much smaller scanning angles can be used since no 'safety margin' needs to be allowed and better signal-to-background ratios can be achieved (p. 232).

(7) The computer can be used to determine the original orientation of the crystal and its accurate unit cell parameters. Ultimately it may be possible to confine the human operator's role to centring the crystal: the computer can then systematically permute the shaft settings until a few reflexions have been found, assign an appropriate unit cell and compute the settings for all other reflexions in turn (p. 51).

More examples could be produced, but it is as yet too early to judge the usefulness of some of the more elaborate suggestions which have been made. So far only a few laboratories have computer-controlled diffractometers and only preliminary reports of their functioning have appeared (e.g. Hamilton, 1964; Cole & Okaya, 1964).

Suitable computers for on-line control

An electronic computer is expensive, especially on the scale of expenditure which X-ray crystallographers have been accustomed to applying to their instrumentation. All the computations required for the operation of a diffractometer, with the possible exception of absorption corrections, are relatively simple: they can be carried out in a very small fraction of the time which must be expended on the measurement of a reflexion in order to accumulate a statistically sufficient number of counts. Consequently, the expensive computer may be idle most of the time. An obvious solution to this situation is provided by the advent of time-sharing computers. These computers can execute several tasks quasi-simultaneously. They execute a particular program, which may be one of control or computation, until they are alerted by an 'interrupt signal' indicating that a particular instrument requires attention. At this point they store the intermediate results of their current operation, attend to a higher priority task, and return to their original program when the necessary action has been taken.

The attractiveness of the time-sharing computer is not quite as great as might appear at first sight. While it is true that the actual computations required for the control of a diffractometer are trivial, closed-loop operation requires the storage of a considerable number of contingency programs. If these programs are stored in the high-speed store of the computer there may be little storage left for other problems. Additional high-speed storage for a computer is expensive, and a computer which contains a backing-up store in the form of a magnetic drum, a magnetic disc file or magnetic tape decks is no longer a particularly small or cheap computer. Storage at several levels can most readily be provided in a very large time-sharing computer: one of the first computer-controlled diffractometers is linked to the I.C.T. *Atlas* computer (Bowden *et al.* 1963). The crystallographer's control over the organization and operation of a very large computer, however, is necessarily minimal: it may be very difficult to provide him with uninterrupted use, even though on a time-sharing basis, of such a computer.

Another difficulty arises in many research laboratories. Much of the work often consists of running new untested programs; a fault in them can result in an overflow into a part of the store which holds diffractometer control programs. The provision of 'hardware lock-out', that is, of circuits which make an accidental overflow impossible, again adds to the complexity and cost of the computer.

It is very difficult to make any firm recommendations as to the most effective type of computer for on-line control of a diffractometer. The four installations listed in Table VI (p. 98) are the only ones about which details have appeared: the first uses a small computer on a non-time-sharing basis, the second and third a large computer on a time-sharing basis, and the last a medium-sized computer shared between a number of diffractometers, but not shared with any general purpose computing. Before the advantages of these systems can be discussed realistically more experience in on-line operation will be needed, not only of single crystal diffractometers, but of scientific measuring instruments of all kinds.

APPENDIX

SUMMARY OF DIFFERENCES BETWEEN X-RAY AND NEUTRON DIFFRACTOMETRY

Property	X-rays	Neutrons	Effect of difference
Flux at specimen	At least 10^{10} quanta/cm²/s, using unmono-chromatized radiation	10^6 to 10^7 neutrons/cm²/s, using radiation from high-flux reactor	Samples are larger with neutrons, and counting times longer
Wavelength λ, and width of wavelength band $\delta\lambda$	λ restricted to characteristic K-radiations of elements between Cr and Ag (~ 2 Å $- 0 \cdot 5$ Å). Characteristic line is very sharp ($\delta\lambda < 10^{-4}$ Å), but superimposed on 'white' background	With a crystal mono-chromator, λ can take any value in the range $0 \cdot 8$ to $2 \cdot 0$ Å; $\delta\lambda$ is between $0 \cdot 02$ and $0 \cdot 05$ Å. With hot and cold sources and mechanical velocity selectors the range of λ can be extended from $0 \cdot 5$ to 10 Å	Long wavelength measurements possible with neutrons. Bragg peaks are wider with neutrons. White background error with X-rays
Nature of general scattering	Electronic: scattering amplitude f falls off with scattering angle	Nuclear: scattering is isotropic	Bragg reflexions can be observed at higher θ values with neutrons. Scattering ampli-tude of atom for neutrons is repre-sented by a single number; for X-rays it must be calculated from the electronic structure and is known only approximately
Absorption coefficient	$\mu \sim 10$ to 10^2	$\mu \sim 10^{-1}$, except for B, Cd, Li and rare-earths	Absorption correc-tion much more important with X-rays

Property	X-rays	Neutrons	Effect of difference
Anomalous dispersion	Always present, even far from an absorption edge. $f = f^0 + \Delta f' + i\Delta f''$ where $\Delta f'$, $\Delta f''$ depend on wavelength	Absent, except for a few elements (e.g. Cd), which have $\Delta f'$, $\Delta f''$ contributions much larger than for X-rays	For most samples $I_{hkl} = I_{\overline{hkl}}$ with neutrons
Extinction	Primary (within each mosaic block) and secondary (between blocks)	Primary extinction less than for X-rays, as scattering amplitude per atomic plane is less; secondary extinction is larger, as beam penetrates further	Especial care required in neutron diffraction to correct for secondary extinction
Thermal diffuse scattering	Takes place with negligible change of wavelength (X-ray energy $\gg$ phonon energy)	Takes place with appreciable change of wavelength (neutron energy $\sim$ phonon energy)	A portion of the one-phonon T.D.S. background of the Bragg reflexions can be removed in neutron diffraction, using an analyser crystal to reflect the diffracted beam
Incoherent background	Compton scattering. Fluorescence scattering (inelastic)	Spin and isotopic incoherent scattering: spin incoherence very prominent with hydrogen (elastic disorder scattering)	Background between reflexions tends to be higher with neutrons, particularly for crystals containing hydrogen
Absolute intensity determination	Very difficult as direct beam cannot be measured easily in presence of white radiation: comparison with reference sample complicated by high absorption	Relatively straightforward	Experimental structure factors are more readily placed on absolute scale with neutrons

Property	X-rays	Neutrons	Effect of difference
Polarization factor p	Varies with θ and is uncertain for crystal-reflected radiation	$p = 1$ for nuclear reflexions	Extinction can be corrected in X-ray diffraction by examining change of intensity with state of polarization
Method of detection	Proportional or scintillation counter, with pulse-height discrimination	BF_3 counter: neutron converted into α-particle	Pulse-height discrimination not possible with neutrons
Velocity of propagation	3×10^{10} cm/s	4×10^5 cm/s (1 Å neutrons); inversely proportional to wavelength	λ can be measured with neutrons by time-of-flight technique. Neutron structure factors can be determined by Laue method

REFERENCES AND AUTHOR INDEX

The numbers in square brackets are the
pages where the references occur.

ABRAHAMS, S. C. (1962). *Rev. scient. Instrum.* **33**, 973. [97]

ABRAHAMS, S. C. (1964). *Acta crystallogr.* **17**, 1190. [296, 298]

ABRAHAMS, S. C. (1965). Private communication. [298].

ABRAHAMS, S. C. *see also* PRINCE & ABRAHAMS (1959).

ABSON, W., SALMON, P. G. & PYRAH, S. (1958*a*). *Proc. Instn elect. Engrs,* **105**B, 357. [141].

ABSON, W., SALMON, P. G. & PYRAH, S. (1958*b*). *Proc. Instn elect. Engrs,* **105**B, 349. [142]

AGRON, P. A. *see* LEVY, AGRON & BUSING (1963).

ALBRECHT, G. (1939). *Rev. scient. Instrum.* **10**, 221. [241]

ALEXANDER, L. E., KUMMER, E. & KLUG, H. P. (1949). *J. appl. Phys.* **20**, 735. [145]

ALEXANDER, L. E. & SMITH, G. S. (1962). *Acta crystallogr.* **15**, 983. [229, 265, 266]

ALEXANDER, L. E. & SMITH, G. S. (1964). *Acta crystallogr.* **17**, 447. [266]

ALIKHANOV, R. A. (1959). *J. exp. theor. Phys. U.S.S.R.* **36**, 1690. [205]

ALLISON, S. K. *see* COMPTON & ALLISON (1935).

ARNDT, U. W. (1949). *J. scient. Instrum.* **26**, 45. [146]

ARNDT, U. W. (1955). *X-ray Diffraction by Polycrystalline Materials,* ch. 7. Eds. H. S. Peiser, H. P. Rooksby and A. J. C. Wilson (London: The Institute of Physics). [106, 107, 122]

ARNDT, U. W. (1963). *Hilger J.* **8**, 4. [82, 98]

ARNDT, U. W. (1964). *Acta crystallogr.* **17**, 1183. [74]

ARNDT, U. W. (1966). *J. scient. Instrum.* To be published. [150]

ARNDT, U. W., COATES, W. A. & RILEY, D. P. (1953). *Proc. phys. Soc. Lond.* B, **66**, 1009. [196]

ARNDT, U. W., FAULKNER, T. H. & PHILLIPS, D. C. (1960). *J. scient. Instrum.* **37**, 68. [72, 89, 90]

ARNDT, U. W., GOSSLING, T. H. & MALLETT, J. F. W. (1966). To be published. [98]

ARNDT, U. W., JONES, F. B. & LONG, A. R. (1965). Unpublished. [72]

ARNDT, U. W. & MACGANDY, E. L. (1962). Unpublished. [82]

ARNDT, U. W., NORTH, A. C. T. & PHILLIPS, D. C. (1964). *J. scient. Instrum.* **41**, 421. [12, 63, 88, 94]

ARNDT, U. W. & PHILLIPS, D. C. (1957). *Acta crystallogr.* **10**, 508. [66]

ARNDT, U. W. & PHILLIPS, D. C. (1959). *Br. J. appl. Phys.* **10**, 116. [75]

ARNDT, U. W. & PHILLIPS, D. C. (1961). *Acta crystallogr.* **14**, 807. [75, 88, 91, 97]

ARNDT, U. W. & PHILLIPS, D. C. (1966). To be published. [94]

ARNDT, U. W. & RILEY, D. R. (1952). *Proc. phys. Soc. Lond.* A, **65**, 74. [180]

316 REFERENCES

ARNDT, U. W. & WILLIS, B. T. M. (1963*a*). *Nucl. Instrum. Meth.* **24**, 155. [69, 82, 98]

ARNDT, U. W. & WILLIS, B. T. M. (1963*b*). *Rev. scient. Instrum.* **34**, 224. [72, 82, 98]

ATKINSON, H. H. (1958). *Phil. Mag.* **3**, 476. [192]

ATOJI, M. (1964). Argonne National Laboratory Report, ANL–6920. [97, 206]

AZAROFF, L. V. (1955). *Acta crystallogr.* **8**, 701. [287]

BACON, G. E. (1962). *Neutron Diffraction* (Oxford: Clarendon Press). [xii, 198, 277, 295]

BACON, G. E. & LOWDE, R. D. (1948). *Acta crystallogr.* **1**, 303. [244]

BACON, G. E., SMITH, J. A. G. & WHITEHEAD, C. D. (1950). *J. scient. Instrum.* **27**, 330. [3]

BAKER, T. W. (1966). A.E.R.E. Harwell Report. [264]

BARNES, D. C. & FRANKS, A. (1962). *J. scient. Instrum.* **39**, 648. [76]

BARRETT, C. S., MUELLER, M. H. & HEATON, L. (1963). *Rev. scient. Instrum.* **34**, 847. [204, 205]

BEAR, R. S. *see* BOLDUAN & BEAR (1949).

BELSON, H. S. (1964). *Rev. scient. Instrum.* **35**, 234. [258]

BENEDICT, T. S. (1955). *Acta crystallogr.* **8**, 747. [67, 97]

BERNAL, J. D. (1926). *Proc. R. Soc.* A, **113**, 117. [2]

BILANIUK, O. M. (1960). *see* BILANIUK, O. M., HAMANN, A. K. & MARSH, B. B. (1960). University of Rochester Report. AT(30–1) 875, May 1960. [150]

BINNS, J. V. (1964). *J. scient. Instrum.* **41**, 715. [76, 93]

BLALOCK, T. V. (1964). *I.E.E. Trans. Nucl. Sci.* NS–11, 365. [156]

BLOCHIN, M.A. (1957). *Physik der Röntgenstrahlen* (Berlin: Verlag d. Technik). [170]

BLOW, D. M. & GOSSLING, T. H. (1965). Unpublished. [306]

BOLDUAN, O. E. A. & BEAR, R. S. (1949). *J. appl. Phys.* **20**, 983. [173]

BOLLINGER, L. M., THOMAS, G. E. & GINTHER, R. J. (1962). *Nucl. Instrum. Meth.* **17**, 97. [144]

BOND, W. L. (1951). *Rev. scient. Instrum.* **22**, 344. [258]

BOND, W. L. (1955). *Acta crystallogr.* **8**, 741. [67, 97]

BOND, W. L. (1960). *Acta crystallogr.* **13**, 814. [263]

BORGONOVI, G. & CAGLIOTI, G. (1962). *Nuovo Cim.* **24**, 1174. [252]

BORRMANN, M. (1941). *Phys. Z.* **42**, 157. [193]

BORRMANN, M. (1950). *Z. Phys.* **127**, 297. [193]

BOSANQUET, C. H. *see* BRAGG, JAMES & BOSANQUET (1921).

BOWDEN, K., EDWARDS, D. & MILLS, O. S. (1963). *Acta crystallogr.* **16**, A 177. [98, 311]

BOWMAN, H. R., HYDE, E. K., THOMPSON, S. G. & JARED, R. C. (1966). *Science*, **151**, 562. [136]

BOZORTH, R. M. & HAWORTH, F. E. (1938). *Phys. Rev.* **53**, 538. [190]

BRAGG, W. H. & BRAGG, W. L. (1913). *Proc. R. Soc.* A, **88**, 428. [1, 97]

BRAGG, W. L., JAMES, R. W. & BOSANQUET, C. H. (1921). *Phil. Mag.* **42**, 1. [248]

BRAGG, W. L. & WEST, J. (1928). *Z. Kristallogr. Miner.* **69**, 118. [288]

BROAD, D. A. G. (1956). *Acta crystallogr.* **9**, 834. [170]

BROCKHOUSE, B. N. (1961). In *Inelastic Scattering of Neutrons in Solids and Liquids*, p. 113. (Vienna: I.A.E.A.) [3]

BROCKMAN, F. G. *see* CORLISS, HASTINGS & BROCKMAN (1953).

BROWN, I. D. (1958). *Acta crystallogr.* **11**, 510. [66]

BROWN, P. J. & FORSYTH, J. B. (1960). *Acta crystallogr.* **13**, 985. [97]

BUDAL, K. (1963). Arkiv f. d. Fysiske Seminar Trondheim (9); *Nucl. Instrum. Meth.* **23**, 132. [150]

BUERGER, M. J. (1942). *X-ray Crystallography* (New York: John Wiley). [xii, 2, 26, 31]

BUERGER, M. J. (1944). *The Photography of the Reciprocal Lattice.* A.S.X.R.E.D. [2]

BUERGER, M. J. (1960). *Crystal Structure Analysis* (New York: John Wiley). [10, 97, 277]

BUERGER, M. J. (1964). *The Precession Method* (New York: John Wiley). [2]

BURAS, B. & LECIEJEWICZ, J. (1964). *Phys. Stat. Sol.* **4**, 349. [217, 218]

BURBANK, R. D. (1964). *Acta crystallogr.* **17**, 434. [222, 229, 265, 266]

BURBANK, R. D. (1965). *Acta crystallogr.* **18**, 88. [291]; **19**, 957. [255].

BUSING, W. R. & LEVY, H. (1957). *Acta crystallogr.* **10**, 180. [241, 243]

BUSING, W. R. *see also* LEVY, AGRON & BUSING (1963).

BUTT, N. M. *see* O'CONNOR & BUTT (1963).

BUTT, N. M. *see* O'CONNOR & BUTT (1965).

BUYERS, T. (1964). Ph.D. Thesis. University of Aberdeen. [291]

CAGLIOTI, G. (1964). *Acta crystallogr.* **17**, 1202. [224, 225]

CAGLIOTI, G. & RICCI, F. P. (1962). *Nucl. Instrum. Meth.* **15**, 155. [211]

CAGLIOTI, G. *see also* BORGONOVI & CAGLIOTI (1962).

CATH, P. G. *see* LADELL & CATH (1963).

CHAMBERS, F. W. *see* COLE, CHAMBERS & DUNN (1962).

CHAMBERS, F. W. *see* COLE, CHAMBERS & WOOD (1961).

CHAMBERS, F. W. *see* COLE, CHAMBERS & WOOD (1962).

CHAMBERS, F. W. *see* COLE, OKAYA & CHAMBERS (1963).

CHANDRASEKHAR, S. (1956). *Acta crystallogr.* **9**, 954. [192, 246]

CHANDRASEKHAR, S. (1960*a*). *Acta crystallogr.* **13**, 588. [192, 246]

CHANDRASEKHAR, S. (1960*b*). *Adv. Phys.* **9**, 363. [192, 246, 247, 284]

CHANDRASEKHAR, S. & PHILLIPS, D. C. (1961). *Nature, Lond.* **190**, 1164. [192]

CHANDRASEKHAR, S. & WEISS, R. J. (1957). *Acta crystallogr.* **10**, 598. [247]

CHANDRASEKARAN, K. S. (1956). *Proc. Indian Acad. Sci.* A **44**, 387. [193]

CHIDAMBARAM, A., SEQUEIRA, A. S. & SIKKA, S. K. (1964). *Nucl. Instrum. Meth.* **26**, 340. [265]

CHIPMAN, D. R. & PASKIN, A. (1958). *J. appl. Phys.* **29**, 1608; **30**, 1992; **30**, 1998. [222]

CLASTRE, J. (1960). *Acta crystallogr.* **13**, 986. [97]

CLIFTON, D. F., FILLER, A. & MCLACHLAN, D. (1951). *Rev. scient. Instrum.* **22**, 1024. [97]

COATES, W. A. *see* ARNDT, COATES & RILEY (1953).

COCEVA, C. (1963). *Nucl. Instrum. Meth.* **21**, 93. [143, 144]

COCHRAN, W. (1950). *Acta crystallogr.* **3**, 268. [2, 97, 146]

COCHRAN, W. (1953). *Acta crystallogr.* **6**, 260. [248

318 REFERENCES

COCHRAN, W. *see also* LIPSON & COCHRAN (1957).

COHEN, L., FRAENKEL, B. S. & KALMAN, Z. H. (1963). *Acta crystallogr.* **16**, 1192. [252]

COLE, D. G. *see* GILLAM & COLE (1953).

COLE, H., CHAMBERS, F. W. & DUNN, H. M. (1962). *Acta crystallogr.* **15**, 138. [252, 254, 255]

COLE, H., CHAMBERS, F. W. & WOOD, C. G. (1961). *J. appl. Phys.* **32**, 1942. [193, 194]

COLE, H., CHAMBERS, F. W. & WOOD, C. G. (1962). *Rev. scient. Instrum.* **33**, 435. [171, 173]

COLE, H. & OKAYA, Y. (1964). ACA Meeting, Paper A 5. [310]

COLE, H., OKAYA, Y. & CHAMBERS, F. W. (1963). *Acta crystallogr.* **16**, A 154; *Rev. scient. Instrum.* **34**, 872. [98]

COLLINS, M. F. *see* LOW & COLLINS (1963).

COMPTON, A. H. & ALLISON, S. K. (1935). *X-rays in Theory and Experiment* (New York: Van Nostrand). [xii, 146, 170, 195, 208]

COOKE-YARBOROUGH, E. H., FLORIDA, C. D. & DAVEY, C. N. (1949). *J. scient. Instrum.* **26**, 124. [123]

CORLISS, L. M., HASTINGS, J. M. & BROCKMAN, F. G. (1953). *Phys. Rev.* **90**, 1013. [3]

COSSLETT, V. E. & NIXON, W. C. (1960). *X-ray Microscopy* (Cambridge University Press). [170, 171]

COWAN, J. P., MACINTYRE, W. M. & THOMAS, R. (1965). ACA Meeting, Paper A 4. [150]

COWAN, J. P., MACINTYRE, W. M. & WERKEMA, G. J. (1963). *Acta crystallogr.* **16**, 221. [233]

CRUICKSHANK, D. W. J. (1956). *Acta crystallogr.* **9**, 757. [288]

CRUICKSHANK, D. W. J. (1956). *Acta crystallogr.* **9**, 747. [295]

CRUICKSHANK, D. W. J. (1960). *Acta crystallogr.* **13**, 774. [296]

DACHS, H. (1961). *Z. Kristallogr. Miner.* **115**, 80. [211]

DARWIN, C. G. (1922). *Phil. Mag.* **43**, 800. [248, 250]

DAVEY, C. N. *see* COOKE-YARBOROUGH, FLORIDA & DAVEY (1949).

DAVIES, D. A., MATHIESON, A. McL. & STIFF, G. M. (1959). *Rev. scient. Instrum.* **30**, 488. [170]

DAWSON, B. (1964*a*). *Acta crystallogr.* **17**, 990. [293]

DAWSON, B. (1964*b*). *Acta crystallogr.* **17**, 997. [293, 294]

DEBYE, P. (1914). *Annln Phys.* **43**, 49. [221]

DEARNALEY, G. & NORTHROP, D. C. (1963). *Semiconductor Counters for Nuclear Radiations* (London: E. and F. N. Spon). [136]

DIAMANT, H. *see* DRENCK, PEPINSKY & DIAMANT (1959).

DRENCK, R., DIAMANT, H. & PEPINSKY, R. (1959). ACA Meeting, Cornell, Paper G 8. [73, 97]

DUMOND, J. W. M. *see* HENKE & DUMOND (1953).

DUMOND, J. W. M. *see* HENKE & DUMOND (1955).

DUNN, H. M. *see* COLE, CHAMBERS & DUNN (1962).

EASTABROOK, J. N. & HUGHES, J. W. (1953). *J. scient. Instrum.* **30**, 317. [146]

EDWARDS, D. W. G. *see* BOWDEN, EDWARDS & MILLS (1963).

EHRENBERG, W. (1949). *J. opt. Soc. Amer.* **39**, 741. [195]

EHRENBERG, W. & FRANKS, A. (1952). *Nature, Lond.* **170**, 1076. [195]

ELLIOTT, A. (1965). *J. scient. Instrum.* **42**, 312. [192, 195]

ELLISON, R. D. *see* LEVY & ELLISON (1960).

EVANS, H. T. (1953). *Rev. scient. Instrum.* **24**, 156. [66, 97]

EVANS, D. F. (1961). *Digital Data* (London: Hilger and Watts). [70]

EVANS, R. C., HIRSCH, P. B. & KELLAR, J. N. (1948). *Acta crystallogr.* **1**, 124. [190, 191]

EWALD, P. P. (1921). *Z. Kristallogr. Miner.* **56**, 129. [2]

FAIRSTEIN, E. (1961a). NAS-NRC Nucl. Sci. Series Rep. **32**, 210. [157]

FAIRSTEIN, E. (1961b). *I.R.E. Trans. Nucl. Sci.* NS-8, 129. [157]

FAIRSTEIN, E. (1962). In *Nuclear Instruments and their Uses*. Ed. A. H. Snell (New York: John Wiley). [164]

FANKUCHEN, I. (1937). *Nature, Lond.* **139**, 193. [190]

FANKUCHEN, I. (1938). *Phys. Rev.* **53**, 910. [190]

FANKUCHEN, I. *see also* WILLIAMSON & FANKUCHEN (1959).

FANKUCHEN, I. *see also* YAKEL & FANKUCHEN (1962).

FARKAS, G. & VARGA, P. (1964). *J. scient. Instrum.* **41**, 704. [129]

FAULKNER, T. H. *see* ARNDT, FAULKNER & PHILLIPS (1960).

FAXÉN, H. (1923). *Z. Phys.* **17**, 266. [221]

FILLER, A. *see* CLIFTON, FILLER & McLACHLAN (1951).

FINCH, J. T. & KLUG, A. (1959). *Nature, Lond.* **183**, 1709. [Plate 1]

FLORIDA, C. D. *see* COOKE-YARBOROUGH, FLORIDA & DAVEY (1949).

FORSYTH, J. B. *see* BROWN & FORSYTH (1960).

FOURNET, G. *see* GUINIER & FOURNET (1955).

FRAENKEL, B. S. *see* COHEN, FRAENKEL & KALMAN (1963).

FRANKS, A. (1955). *Proc. phys. Soc.* B, **68**, 1054. [195]

FRANKS, A. *see also* BARNES & FRANKS (1962).

FRANKS, A. *see also* EHRENBERG & FRANKS (1952).

FRIEDRICH, W., KNIPPING, P. & VON LAUE, M. (1912). *Proc. Bavarian Acad. Sci.* p. 303. Reprinted in *Naturwissenschaften* (1952), p. 368. [1]

FURNAS, T. C. (1957). *Single Crystal Orienter Instruction Manual* (Milwaukee: The General Electric X-ray Corp.). [97, 171, 241, 257, 260]

FURNAS, T. C. & HARKER, D. (1955). *Rev. scient. Instrum.* **26**, 449. [39, 51, 79, 97]

GELLER, S. & KATZ, H. (1962). *Bell Syst. tech. J.* **41**, 425. [67]

GETTING, I. A. (1947). *Phys. Rev.* **53**, 103. [123]

GILLAM, E. & COLE, D. G. (1953). *Phil. Mag.* **44**, 999. [196]

GILLESPIE, A. B. (1953). *Signal, Noise and Resolution in Nuclear Counter Amplifiers* (London: Pergamon Press). [114, 164]

GINTHER, R. J. *see* BOLLINGER, THOMAS & GINTHER (1962).

GOODRICH, G. W. & WILEY, W. C. (1962). *Rev. scient. Instrum.* **33**, 761. [130, 131]

GOSSLING, T. H. *see* ARNDT, GOSSLING & MALLETT (1966).

GOSSLING, T. H. *see* BLOW & GOSSLING (1965).

GOULDING, F. S. & HANSEN, W. L. (1961). *Nucl. Instrum. Meth.* **12**, 249. [136, 165]

GOULDING, F. S. *see* LANDIS & GOULDING (1964).

GUILD, J. (1956). *The Interference Systems of Cross Diffraction Gratings: Theory of Moiré Fringes* (Oxford: Clarendon Press). [71]

320 REFERENCES

GUINIER, A. & FOURNET, G. (1955). *Small-Angle Scattering of X-rays* (New York: John Wiley). [171]
HAMILTON, W. C. (1957). *Acta crystallogr.* **10**, 629. [244, 248, 250]
HAMILTON, W. C. (1963). *Acta crystallogr.* **16**, 609. [244, 250]
HAMILTON, W. C. (1964). ACA Meeting, Bozeman, Paper A 6. [98, 310]
HANSEN, W. L. *see* GOULDING & HANSEN (1961).
HARGREAVES, C. M., PRINCE, E. & WOOSTER, W. A. (1952). *J. scient. Instrum.* **29**, 82. [196]
HARKER, D. *see* FURNAS & HARKER (1955).
HARRIS, D. H. C. (1961a). A.E.R.E. Harwell Report, R-3688. [144]
HARRIS, D. H. C. (1961b). A.E.R.E. Harwell Report, M-827. [144]
HASTINGS, J. M. *see* CORLISS, HASTINGS & BROCKMAN (1953).
HAWORTH, F. E. *see* BOZORTH & HAWORTH (1938).
HEATON, L. & MUELLER, M. H. (1960). Private communication; see *Rev. scient. Instrum.* (1961), **31**, 456; *ibid.* (1963), **34**, 74. [72]
HEATON, L. *see* BARRETT, MUELLER & HEATON (1963).
HENKE, B. & DUMOND, J. W. M. (1953). *Phys. Rev.* **89**, 1300. [195]
HENKE, B. & DUMOND, J. W. M. (1955). *J. appl. Phys.* **26**, 903. [195]
HIRSCH, P. B. (1955). *X-ray Diffraction by Polycrystalline Materials*, ch. 11. Eds. H. S. Peiser, H. P. Rooksby and A. J. C. Wilson (London: The Institute of Physics). [171, 173, 179]
HIRSCH, P. B. *see also* EVANS, HIRSCH & KELLAR (1948).
HODGSON, L. I. *see* BOWDEN, K. *et al.* (1963).
HOFSTADTER, R., O'DELL, E. W. & SCHMIDT, C. T. (1964). *Rev. scient. Instrum.* **35**, 246. [125]
HOFSTADTER, R. *see also* WEST, MAYERHOF & HOFSTADTER (1951).
HOLMES, K. C. (1964). Private communication. [196]
HOSOYA, S. *see* MIYAKE, TOGAWA & HOSOYA (1964).
HUGHES, D. J. (1953). *Pile Neutron Research* (Cambridge, Mass.: Addison-Wesley). [xii, 215]
HUGHES, J. W. *see* EASTABROOK & HUGHES (1953).
HURST, D. G., PRESSESSKY, A. J. & TUNNICLIFFE, P. R. (1950). *Rev. scient. Instrum.* **21**, 705. [3]
HUXLEY, H. E. (1953). *Acta crystallogr.* **6**, 457. [173]
HYDE, E. K. *see* BOWMAN, HYDE, THOMPSON & JARED (1966).
International Tables. (*The International Tables for X-ray Crystallography.*) 3 vols. (Birmingham: Kynoch Press). [*passim*]
INTERNATIONAL UNION CRYSTALLOGRAPHIC APPARATUS COMMISSION (1956). *Acta crystallogr.* **9**, 976. [79]
JAMES, R. W. (1962). *The Optical Principles of the Diffraction of X-rays* (London: G. Bell). [16, 244, 245, 292]
JAMES, R. W. *see also* BRAGG, JAMES & BOSANQUET (1921).
JARED, R. C. *see* BOWMAN, HYDE, THOMPSON & JARED (1966).
JEFFERY, J. W. & ROSE, K. M. (1964). *Acta crystallogr.* **17**, 343. [238, 239]
JENTZSCH, F. & NÄHRING, E. (1931). *Z. tech. Phys.* **12**, 185. [179]
JOHANN, H. H. (1931). *Z. Phys.* **69**, 185. [190]
JOHANSSON, T. (1932). *Naturwissenschaften*, **20**, 758. [191]
JONES, F. B. *see* ARNDT, JONES & LONG (1965).
KALMAN, Z. H. *see* COHEN, FRAENKEL & KALMAN (1963).

KARTHA, G. (1953). *Acta crystallogr.* 6, 817. [289]

KATZ, H. *see* GELLER & KATZ (1962).

KELLAR, J. N. *see* EVANS, HIRSCH & KELLAR (1948).

KEVEY, A. (1964). Brookhaven Report, BNL 851 (T-337). [200]

KIRKPATRICK, P. (1939). *Rev. scient. Instrum.* 10, 186. [184]

KIRKPATRICK, P. (1944). *Rev. scient. Instrum.* 15, 223. [184]

KLUG, A. *see* FINCH & KLUG (1959).

KLUG, H. P. *see* ALEXANDER, KUMMER & KLUG (1949).

KNIPPING, P. *see* FRIEDRICH, KNIPPING & VON LAUE (1912).

KOHLER, T. R. *see* PARRISH & KOHLER (1956).

KUMMER, E. *see* ALEXANDER, KUMMER & KLUG (1949).

LADELL, J. & CATH, P. G. (1963). Norelco Reporter Special Issue. [72, 97]

LADELL, J. & LOWITSCH, K. (1960). *Acta crystallogr.* 13, 205. [75]

LADELL, J., PARRISH, W. & SPIELBERG, N. (1963). ACA Conference Report. *See* LADELL & CATH (1963). [75]

LANDIS, D. & GOULDING, F. S. (1964). Publication 1184. Nat. Acad. Sci., N.R.C., Washington, D.C. [130, 159]

LANG, A. R. (1954). *Rev. scient. Instrum.* 25, 1039. [21]

LAUE, M. VON *see* FRIEDRICH, KNIPPING & VON LAUE (1912).

LAUE, M. VON (1960). *Röntgenstrahlinterferenzen* (Frankfurt a. M.: Akademische Verlagsgesellschaft). [193]

LAVAL, J. (1938). *C.r. hebd. Séanc. Acad. Sci.*, Paris, 207, 169. [221]

LECIEJEWICZ, J. *see* BURAS & LECIEJEWICZ (1964).

LEVY, H. A., AGRON, P. A. & BUSING, W. R. (1963). ACA Meeting, Cambridge, Mass., Paper E 7, *see* BUSING, SMITH, PETERSON & LEVY (1964). [98]

LEVY, H. A. & ELLISON, R. D. (1960). *Acta crystallogr.* 13, 270. [287]

LEVY, H. A. *see* BUSING & LEVY (1957).

LIPSON, H. & COCHRAN, W. (1957). *The Determination of Crystal Structures* (London: Bell). [288]

LIPSON, H., NELSON, J. B. & RILEY, D. P. (1945). *J. scient. Instrum.* 22, 184. [189]

LONG, A. R. *see* ARNDT, JONES & LONG (1965).

LONGLEY, W. (1963). Ph.D. Thesis, University of London. [173]

LONSDALE, K. (1947). *Min. Mag.*, Lond. 28, 14. [251]

LONSDALE, K. (1948). *Acta crystallogr.* 1, 12. [2]

LOW, G. C. E. & COLLINS, M. F. (1963). *J. appl. Phys.* 34, 1195. [216]

LOWDE, R. D. (1950). *Rev. scient. Instrum.* 21, 835. [142]

LOWDE, R. D. (1956). *Acta crystallogr.* 9, 151. [217]

LOWDE, R. D. *see also* BACON & LOWDE (1948).

LOWITZSCH, K. *see* LADELL & LOWITZSCH (1960).

LUTZ, G. (1960). *Kerntechnik*, 2, 391. [97]

MACGANDY, E. L. *see* ARNDT & MACGANDY (1962).

MACINTYRE, W. M. (1963). *Acta crystallogr.* 16, 315. [289]

MACINTYRE, W. M. *see also* COWAN, MACINTYRE & THOMAS (1965).

MACINTYRE, W. M. *see also* COWAN, MACINTYRE & WERKEMA (1963).

MACK, M. & SPIELBERG, N. (1958). *Spectrochim. Acta*, 12, 169. [271]

MCLACHLAN, D. *see* CLIFTON, FILLER & MCLACHLAN (1951).

McReynolds, A. W. (1952). *Phys. Rev.* **88**, 958. [205]

Mallett, J. F. W. (1963). A.E.R.E. Harwell Memorandum. [143]

Mallett, J. F. W. *see also* Arndt, Gossling & Mallett (1966).

Malmberg, P. R. *see* Sun, Malmberg & Pecjak (1956).

Martin, A. J. P. *see* Wooster & Martin (1936).

Mathews, F. S. *see* North, Phillips & Mathews (1966).

Mathieson, A. McL. (1958). *Acta crystallogr.* **11**, 433. [75, 97]

Mathieson, A. McL. *see* Davies, Mathieson & Stiff (1959).

Mayer, E. A. (1964). *Z. analyt. Chem.* **205**, 153. [51, 98]

Mayerhof, W. E. *see* West, Mayerhof & Hofstadter (1951).

Mills, O. S. *see* Bowden, Edwards & Mills (1963).

Miyake, S., Togawa, S. & Hosoya, S. (1964). *Acta crystallogr.* **17**, 1083. [189, 286]

Moon, R. M. & Shull, C. G. (1964). *Acta crystallogr.* **17**, 805. [252, 254, 255]

Mueller, M. H. *see* Barrett, Mueller & Heaton (1963).

Mueller, M. H. *see also* Heaton & Mueller (1960).

Nähring, E. *see* Jentzsch & Nähring (1931).

Nelson, J. B. *see* Lipson, Nelson & Riley (1945).

Nicholson, K. P. & Snelling, G. F. (1955). *Br. J. appl. Phys.* **6**, 104. [144]

Nilsson, N. (1957). *Ark. Fys.* **12**, 247. [222, 226]

Nixon, W. C. *see* Cosslett & Nixon (1960).

North, A. C. T. (1964). *J. scient. Instrum.* **41**, 41. [306]

North, A. C. T., Phillips, D. C. & Mathews, F. S. (1966). To be published. [241, 242]

North, A. C. T. *see* Arndt, North & Phillips (1964).

Northrop, D. C. *see* Dearnaley & Northrop (1963).

O'Connor, D. A. & Butt, N. M. (1963). *Physics Letters,* **7**, 233. [222]

O'Connor, D. A. & Butt, N. M. (1965). Private communication. [223]

O'Dell, E. W. *see* Hofstadter, O'Dell & Schmidt (1964).

Okaya, Y. *see* Cole, Okaya & Chambers (1963).

Okaya, Y. *see* Cole & Okaya (1964).

Otnes, K. & Palevsky, H. (1963). In *Inelastic Scattering of Neutrons in Solids and Liquids,* vol. 1, p. 95. (Vienna: I.A.E.A.) [215]

Owen, R. B. (1958). *I.R.E. Trans. Nucl. Sci.* NS-5, p. 189. [159]

Owen, R. B. (1961). *5th Int. Instrum. Meas. Conf., Stockholm,* IMS 10. [159]

Parrish, W. (1962). *Advances in X-ray Diffractometry and X-ray Spectrography* (Eindhoven: Centrex). [171]

Parrish, W. & Kohler, T. R. (1956). *Rev. scient. Instrum.* **27**, 795. [109, 118, 181]

Parrish, W. & Spielberg, N. (1964). ACA Meeting. Paper B 5. [171]

Parrish, W. *see also* Ladell, Parrish & Spielberg (1963).

Parrish, W. *see also* Roberts & Parrish (1962).

Paskin, A. *see* Chipman & Paskin (1958).

Pecjak, F. A. *see* Sun, Malmberg & Pecjak (1956).

Palevsky, H. *see* Otnes & Palevsky (1962).

Pell, E. M. (1960). *J. appl. Phys.* **31**, 291. [135]

PEPINSKY, R. *see* DRENCK, PEPINSKY & DIAMANT (1959).

PHILLIPS, D. C. (1964). *J. scient. Instrum.* **41**, 123. [12, 52, 56, 59]

PHILLIPS, D. C. (1966). Unpublished. [120]

PHILLIPS, D. C. *see also* ARNDT, FAULKNER & PHILLIPS (1960).

PHILLIPS, D. C. *see also* ARNDT, NORTH & PHILLIPS (1964).

PHILLIPS, D. C. *see also* ARNDT & PHILLIPS (1957).

PHILLIPS, D. C. *see also* ARNDT & PHILLIPS (1959).

PHILLIPS, D. C. *see also* ARNDT & PHILLIPS (1961).

PHILLIPS, D. C. *see also* ARNDT & PHILLIPS (1966).

PHILLIPS, D. C. *see also* NORTH, PHILLIPS & MATHEWS (1966).

PHILLIPS, F. C. (1963). *An Introduction to Crystallography*, 3rd. ed. (London: Longmans). [xii]

POTTER, R. (1962). *J. scient. Instrum.* **37**, 379. [76, 97]

POWELL, M. J. D. (1966). A.E.R.E. Harwell Report. [51, 255]

PRESSESSKY, A. J. *see* HURST, PRESSESSKY & TUNNICLIFFE (1950).

PRESSMAN, A. I. (1959). *Design of Transistorized Circuits for Digital Computers* (New York: John F. Rider). [154]

PREWITT, C. T. (1960). *Z. Kristallogr. Miner.* **114**, 355. [26]

PRICE, W. J. (1964). *Nuclear Radiation Detection*, 2nd ed. (New York: Magraw Hill). [154]

PRINCE, E. *see* HARGREAVES, PRINCE & WOOSTER (1952).

PRINCE, E. & ABRAHAMS, S. C. (1959). *Rev. scient. Instrum.* **30**, 581. [98]

PYRAH, S. *see* ABSON, SALMON & PYRAH (1958a).

PYRAH, S. *see* ABSON, SALMON & PYRAH (1958b).

RADEKA, V. (1963). *I.E.E. Trans. Nuclear Sci.* N.S.-11, 358. [155]

RAMASESHAN, S. (1964). In *Advanced Methods of Crystallography* (London: Academic Press). [260]

RENNINGER, M. (1937). *Z. Phys.* **106**, 141. [252]

RICCI, F. P. *see* CAGLIOTI & RICCI (1962).

RILEY, D. P. *see* ARNDT, COATES & RILEY (1953).

RILEY, D. R. *see* ARNDT & RILEY (1952).

RILEY, D. P. *see* LIPSON, NELSON & RILEY (1945).

ROBERTS, B. W. & PARRISH, W. (1962). *Int. Tables*, vol. 3, §2.3. [182, 188, 189, 192]

ROSE, K. M. *see* JEFFERY & ROSE (1964).

ROSS, P. A. (1928). *J. opt. Soc. Am.* **16**, 375, 433. [182]

ROUSE, K. & WILLIS, B. T. M. (1966). A.E.R.E. Harwell Report. [299]

SALMON, P. G. *see* ABSON, SALMON & PYRAH (1958a).

SALMON, P. G. *see* ABSON, SALMON & PYRAH (1958b).

SANDOR, E. (1964). Private communication. [192]

SANTORO, A. & ZOCCHI, M. (1964). *Acta crystallogr.* **17**, 597. [255]

SCHMIDT, C. T. *see* HOFSTADTER, O'DELL & SCHMIDT (1964).

SCHWARTZ, L. H. (1965). Private communication. [218, 219]

SEQUEIRA, A. S. *see* CHIDAMBARAM, SEQUEIRA & SIKKA (1964).

SHULL, C. G. (1960). MIT Report AFOSR TR 60-111. [203, 205]

SHULL, C. G. & WOLLAN, E. O. (1951). *Phys. Rev.* **81**, 527. [205]

SHULL, C. G. *see also* MOON & SHULL (1964).

SHULL, C. G. *see* WOLLAN & SHULL (1948).

SIKKA, S. K. *see* CHIDAMBARAM, SEQUEIRA & SIKKA (1964).

SMALLMAN, R. E. *see* WILLIAMSON & SMALLMAN (1953).
SMITH, G. S. *see* ALEXANDER & SMITH (1962).
SMITH, G. S. *see* ALEXANDER & SMITH (1964).
SMITH, J. A. G. *see* BACON, SMITH & WHITEHEAD (1950).
SNELLING, G. F. *see* NICHOLSON & SNELLING (1955).
SPIELBERG, N. *see* LADELL, PARRISH & SPIELBERG (1963).
SPIELBERG, N. *see* MACK & SPIELBERG (1958).
SPIELBERG, N. *see* PARRISH & SPIELBERG (1964).
STIFF, G. M. *see* DAVIES, MATHIESON & STIFF (1959).
STURM, W. J. (1947). *Phys. Rev.* **71**, 757. [205]
SUN, K. H., MALMBERG, P. R. & PECJAK, F. A. (1956). *Nucleonics*, **14** (July), 46.
SZABO, P. (1961). *Acta crystallogr.* **14**, 1206. [247]
TAYLOR, A. (1949). *J. scient. Instrum.* **26**, 225. [170]
TEARE, P. W. (1960). *J. scient. Instrum.* **37**, 132. [189]
THOMAS, G. E. *see* BOLLINGER, THOMAS & GINTHER (1962).
THOMAS, R. *see* COWAN, MACINTYRE & THOMAS (1965).
THOMPSON, S. G. *see* BOWMAN, HYDE, THOMPSON & JARED (1966).
TOGAWA, S. *see* MIYAKE, TOGAWA & HOSOYA (1964).
TUNNICLIFFE, P. R. *see* HURST, PRESSESSKY & TUNNICLIFFE (1950).
VARGA, P. *see* FARKAS & VARGA (1964).
WAGNER, B., WITTE, H. & WÖLFEL, E. (1955). *Z. phys. Chem.* (N.F.), **3**, 273. [291]
WALLER, I. (1928). *Z. Phys.* **51**, 213. [221]
WASER, J. (1951). *Rev. scient. Instrum.* **22**, 563. [23]
WATSON, H. C. (1965). Private communication. [11]
WELLS, M. (1960). *Acta crystallogr.* **13**, 722. [241]
WEISSENBERG, K. (1924). *Z. Phys.* **23**, 229. [2]
WERKEMA, G. J. *see* COWAN, MACINTYRE & WERKEMA (1963).
WEST, H. I., MAYERHOF, W. E. & HOFSTADTER, R. (1951). *Phys. Rev.* **81**, 141. [123]
WEST, J. *see* BRAGG & WEST (1928).
WESTCOTT, C. H. (1948). *Proc. R. Soc.* A, **194**, 508. [145]
WHITEHEAD, C. D. *see* BACON, SMITH & WHITEHEAD (1950).
WILEY, W. C. *see* GOODRICH & WILEY (1962).
WILKINSON, D. H. (1950). *Ionization Chambers and Counters* (Cambridge University Press). [107, 110, 112, 113]
WILLIAMSON, D. T. N. (1960). *Progress in Automation*, **1**, 127. [71]
WILLIAMSON, G. K. & SMALLMAN, R. E. (1953). *J. scient. Instrum.* **30**, 341. [196]
WILLIAMSON, R. S. & FANKUCHEN, I. (1959). *Rev. scient. Instrum.* **30**, 908. [190]
WILLIS, B. T. M. (1960). *Acta crystallogr.* **13**, 763. [208, 211]
WILLIS, B. T. M. (1962a). *Br. J. appl. Phys.* **13**, 548. [40]
WILLIS, B. T. M. (1962b). *J. scient. Instrum.* **39**, 590. [213]
WILLIS, B. T. M. (1962c). In *Pile Neutron Research in Physics*, p. 455. (Vienna: I.A.E.A.) [248, 249]
WILLIS, B. T. M. (1963). *Proc. R. Soc.* A, **274**, 122. [245, 252]
WILLIS, B. T. M. (1965). *Acta crystallogr.* **18**, 75. [296]

WILLIS, B. T. M. *see also* ARNDT & WILLIS (1963*a*).
WILLIS, B. T. M. *see also* ARNDT & WILLIS (1963*b*).
WILLIS, B. T. M. *see also* ROUSE & WILLIS (1965).
WILSON, A. J. C. (1942). *Nature, Lond.* **150**, 152. [289]
WITTE, H. *see* WAGNER, WITTE & WÖLFEL (1955).
WITTE, H. & WÖLFEL, E. (1955). *Z. phys. Chem.* **3**, 296. [248]
WÖLFEL, E. *see* WAGNER, WITTE & WÖLFEL (1955).
WÖLFEL, E. *see also* WITTE & WÖLFEL (1955).
WOLLAN, E. O. & SHULL, C. G. (1948). *Phys. Rev.* **73**, 830. [3, 205]
WOLLAN, E. O. *see also* SHULL &WOLLAN (1951).
WOOD, C. G. *see* COLE, CHAMBERS & WOOD (1961).
WOOD, C. G. *see* COLE, CHAMBERS & WOOD (1962).
WOOSTER, W. A. (1965). *J. scient. Instrum.* **42**, 685. [51, 260]
WOOSTER, W. A. & MARTIN, A. J. P. (1936). *Proc. R. Soc.* A, **155**, 150. [98]
WOOSTER, W. A. & WOOSTER, A. M. (1962). *J. scient. Instrum.* **39**, 103. [47, 98]
WOOSTER, W. A. *see also* HARGREAVES, PRINCE & WOOSTER (1952).
WUENSCH, B. J. (1963). *Acta crystallogr.* **16**, 1259. [298, 299]
YAKEL, H. L. & FANKUCHEN, I. (1962). *Acta crystallogr.* **15**, 1188. [32]
YOUNG, R. A. (1963). *Z. Kristallogr. Miner.* **118**, 233. [184 ff.]
ZACHARIASEN, W. H. (1945). *Theory of X-ray Diffraction in Crystals* (New York: John Wiley). [244, 245]
ZACHARIASEN, W. H. (1963). *Acta crystallogr.* **16**, 1139. [250]
ZACHARIASEN, W. H. *Acta crystallogr.* **18**, 705. [252]
ZINN, W. H. (1947). *Phys. Rev.* **71**, 752. [3]
ZOCCHI, M. *see* SANTORO & ZOCCHI (1964).

SUBJECT INDEX